T0321370

Excitonic and Vibrational Dynamics in Nanotechnology

Quantum Dots vs. Nanotubes

Excitonic and Vibrational Dynamics in Nanotechnology

Quantum Dots vs. Nanotubes

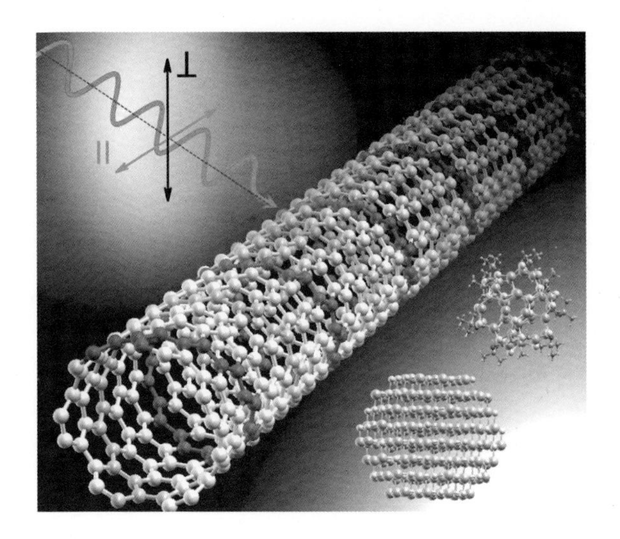

Svetlana V Kilina
CINT, Los Alamos National Laboratory, USA

Bradley F Habenicht
University of Washington, USA

Pan Stanford Publishing

Published by

Pan Stanford Publishing Pte. Ltd.
5 Toh Tuck Link
Singapore 596224

Distributed by

World Scientific Publishing Co. Pte. Ltd.

5 Toh Tuck Link, Singapore 596224

USA office: 27 Warren Street, Suite 401-402, Hackensack, NJ 07601

UK office: 57 Shelton Street, Covent Garden, London WC2H 9HE

British Library Cataloguing-in-Publication Data
A catalogue record for this book is available from the British Library.

EXCITONIC AND VIBRATIONAL DYNAMICS IN NANOTECHNOLOGY
Quantum Dots vs. Nanotubes

ISBN-13 978-981-4241-30-4
ISBN-10 981-4241-30-X

Printed in Singapore by B & JO Enterprise

For my high school physics teacher,
Tamara Petrovna Boikocheva, the first person who opened
the door for me to the wonderful world of physics
and scientific research

Preface

Electronic structure and relaxation of photoexcitations in nano-size materials, such as carbon nanotubes (CNTs) and quantum dots (QDs), are strongly affected by interaction with phonons. We use a wide arsenal of theoretical tools to approach these phenomena. In order to investigate the intricate details of phonon-induced dynamics in QDs, this study uses a novel state-of-the-art quantum-classical approach that combines a molecular dynamics formalism based on density functional theory with a trajectory surface hopping approach. Applying this method, we calculate electronic structure and real-time atomistic relaxation dynamics of charge carriers in QDs made from different materials, such as PbSe and CdSe, that are widely used in nanotechnology. While the sizes of the systems we study are close in size to the typical experimentally investigated QDs, our numeric calculations have an accuracy similar to first-principle quantum mechanical methods. Using this approach, we provide information about the mechanisms that occur on the atomic level and that are extremely difficult – if not impossible – to probe experimentally. We specifically focus on the phonon bottleneck effect in strongly confined QDs – a mysterious phenomenon predicted theoretically but not observed experimentally. We show that PbSe and CdSe have drastically different electronic band structures. Despite this difference, both QDs demonstrate fast subpicosecond relaxation and the absence of the phonon bottleneck, which agree with experiments. We present two rationalizations for such fast relaxation. First, a surface reconstruction and the deviation from the absolute spherical symmetry of the QD lead to a dense distribution of electronic states near the band edges. Most of these states are optically dark; however, they can still help the relaxation process avoiding the phonon bottleneck. Second, localization of wave functions and strong nonadiabatic electron-phonon coupling in small QDs both enhance the probability of multiphonon processes opening a new channel of relaxation and increasing relaxation rates.

In the second part of the thesis we address excited state phenomena in CNTs using the excited-state molecular dynamics methodology that is based on the time-dependent Hartree-Fock approach. This method incorporates electron-hole interactions (excitonic effects), which are essential in CNTs, and makes simulations of exciton-vibrational dynamics in very large systems (up to one thousand atoms in size) possible, while retaining the necessary quantitative accuracy. Based on this approach, we analyze in detail the nature of the strongly bound first and second excitons in CNTs for a number of different tubes, emphasizing emerging size-scaling laws. Characteristic delocalization properties of excited states are identified by the underlying photoinduced changes in charge densities and bond orders. We also estimate the exciton-phonon coupling and its size-scaling law in different CNTs by calculating Huang-Rhys factors, vibrational relaxation and Stokes shift energies, which increase with increasing tube diameter. Due to the rigid structure, exciton-phonon coupling is much weaker in SWCNTs compared to QDs and to typical molecular materials. Yet in the ground state, a CNT surface experiences the corrugation associated with electron-phonon interactions. Vibrational relaxation following photoexcitation reduces this corrugation, leading to a local distortion of the tube surface, which is similar to the formation of self-trapped excitons in conjugated polymers.

Both numerical approaches provide observables, such as relaxation rates in QDs and Huang-Rhys factors and Stokes shift energies in CNTs, that are possible to detect experimentally. Thus, our results allow for better understanding of photoinduced electronic dynamics in nanomaterials, guiding design of new experimental probes, and, potentially may lead to new nanotechnological applications.

Svetlana Kilina
Los Alamos National Laboratory

Contents

Chapter 1

Introduction

Rapid advances in chemical synthesis and fabrication techniques have led to novel nano-sized materials that exhibit unique and often unforeseen properties. One of the greatest advantages of these nanosystems is the ability to control their electronic and optical properties through the sample's size, shape, and topology. This flexibility is ideal for application in several fields ranging from electronics and optelectronics to biology and medicine. The design of nanoelectronic devices requires a clear understanding of the fundamental properties of nanomaterials. When nanomaterials absorb a light quantum, two charged particles are created simultaneously, an electron and a hole, or an exciton, in a process called photoexcitation. The electronic structure changes depending on how strongly the optically excited electrons and holes are bound together by Coulomb forces (excitonic effects) and how sensitive they are to the structure and geometry of the atomic lattice (phonon or vibrational effects). Knowledge of photoinduced charge carrier dynamics in nanomaterials, i.e. the evolution of the physical process that electron-hole and electron-phonon interactions bring about, will help to achieve an effective functionality of prospective nanoelectronic devices.

Semiconductor quantum dots (QDs) and single-walled carbon nanotubes (SWCNTs) are two of the most promising examples of low-dimensional nanomaterials. In recent decades they have been the subject of many experimental, theoretical, and technological investigations . Despite this intensive study, the phonon-mediated dynamics upon photoexcitation and the excitonic (interacting electron-hole pairs) effects in these nanosystems, are still not well-understood.

In this work we investigate QDs and SWCNTs using quantum-chemical calculations that describe excited state phenomena in detail. This method provides information about atomic-level mechanisms that are extremely difficult – if not impossible – to probe experimentally.

1.1 Common Features of Low-Dimensional Nanomaterials

Both QDs and SWCNTs can be considered nanoscale derivatives of bulk materials. For example, a QD, or nanocrystal, is made from an inorganic semiconductor crystal a few nanometers in size. While QDs typically have a roughly spherical shape, cylindrical (nano-rods) and rectangular shapes are also known. Because, in absence of reconstruction, the material remains structurally identical to the bulk crystal at the atomic level, the electronic structure of QDs made from different host materials also differ in accordance with the electronic structure of the parent bulk crystal. The nanometer scale of a QD leads to its 0-dimensionality (0-D). As a result the motion of charge carriers is confined in all three spatial directions. This confinement is caused by electrostatic potentials imposed by the surface or by the presence of an interface between different semiconductor materials (e.g. in core-shell nanocrystal systems) [Klimov (2004)].

Analogously, SWCNTs can be considered quasi-one-dimensional (1-D) derivatives of bulk graphite. A one-atom-thick layer (graphene) is rolled into a long rigid cylinder a few nanometers in diameter. The SWCNT circumference is expressed by a chirality vector C_h connecting two crystallographically equivalent sites of the graphene sheet: $\vec{C}_h = m\vec{a}_1 + n\vec{a}_2$, where $\vec{a}_{1,2}$ are the unit vectors of the hexagonal honeycomb graphene lattice (for illustration, see Fig. 1.1). The structure of any SWCNT can therefore be uniquely described by a pair of integers (n,m) that defines its chiral vector [Saito *et al.* (1998a)]. Similar to bulk graphite SWCNTs preserve an extended backbone structure that supports a delocalized mobile π-electron system. However, unlike graphite the quasi 1-D geometry of SWCNTs confines the motion of electrons or holes in two spatial directions and allows free propagation in the third one along the tube length. Therefore, in going from graphene to a SWCNT by folding one has to account for the additional quantization arising from electron confinement around the tube circumference. This circumferential component of the wave vector k_C can only take values fulfilling the condition $k_C.C_h = 2\pi N$, where N is an integer [Saito *et al.* (1998a)]. As a result each graphene band splits into a number of one-dimensional subbands labeled by N. These allowed energy states are cuts of the graphene band structure. When these cuts pass through a K point (Fermi point) of the graphene Brillouin zone the tube is metallici; other-

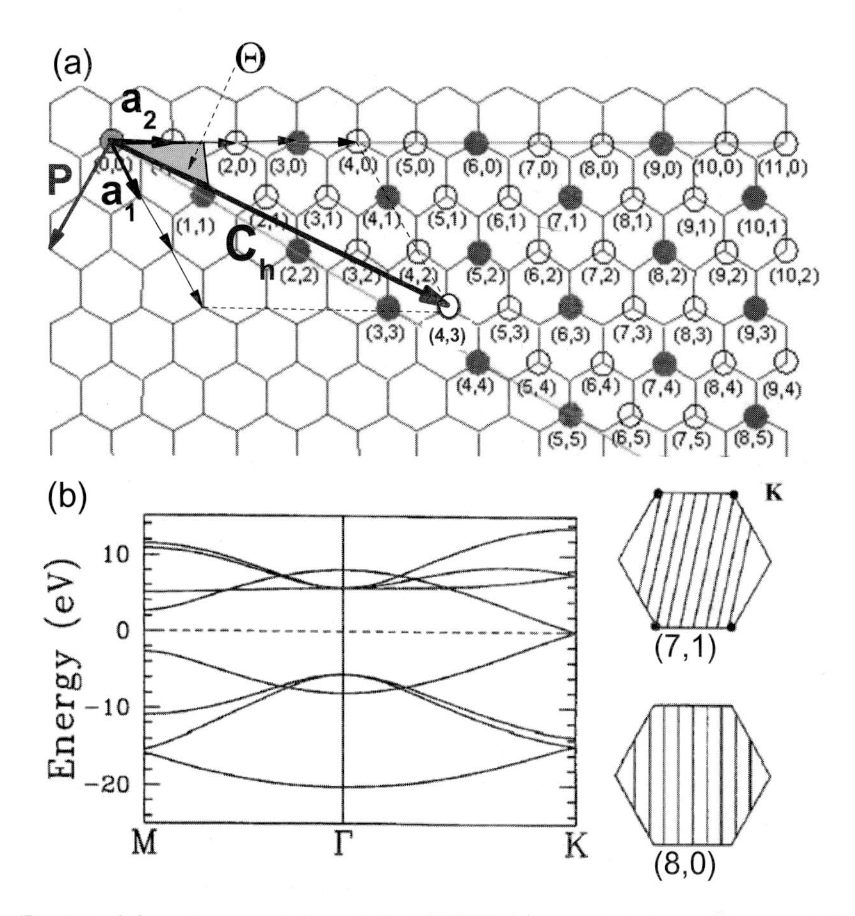

Fig. 1.1 Schematic presentation of the SWCNT (a) geometrical parameters and (b) band structure through the folding of a graphene strip. a) Vectors $\mathbf{a_1}$ and $\mathbf{a_2}$ are the unit vectors of the hexagonal graphene lattice. The chirality vector $\mathbf{C_h} = n\mathbf{a_1} + m\mathbf{a_2}$ (blue arrow), chiral angle Θ, and translational vector P directed along the tube length (purple arrow) of a (4,3) SWCNT are shown as an example. If the ends of the C_h vector are connected, one obtains a tube (4,3). The tube circumference is defined by C_h and the tube axis is directed perpendicular to the chiral vector and parallel to the translational vector P. Numbers in round brackets stand for tube indices (n,m). b) The band diagram (E-k) of a SWCNT calculated by tight-binding model [Saito *et al.* (1998a)]: vector k_P corresponding to translational vector P. Each band relates to one of the N quantized wave vectors in the k_C direction which corresponds to the C_h-vector. Here the Fermi energy is chosen to be 0. The panel on the right shows the reciprocal lattices of (7,1) and (8,0) SWCNT obtained from the graphene reciprocal lattice. Essentially one can imagine N cuts of the graphene dispersion relation. Each cut corresponds to a quantized wave vector in the k_C direction folded back into the first Brillouin zone to result in an E-k plot with N discrete energy bands. If for a particular (n,m) nanotube the cross-section passes through a K point in the 2D Brillouin zone of graphene, then the 1D energy bands of the nanotube have no bandgap and will therefore be metallic. Such tubes are marked by red dots in the plot. (7,1) is an example of a metalic SWCNT. If the cross-sections do not pass through a K point, then the nanotube is semiconducting with a finite bandgap as in (8,0) SWCNT. For color reference, turn to page 149.

wise the tube is semiconducting. It can be shown that an (n,m) SWCNT is metallic when $n = m$ (armchair tubes). It has a small gap when $n - m = 3i$ where i is an integer. CNTs with $n - m \neq 3i$ are truly semiconducting as in the case of most tubes with m=0 called zigzag SWCNTs. Figure 1.1 summarizes the above discussion. Thus, the diameter and direction of the tubes rolling (C_h-vector) determine the main optelectronic characteristics of a SWCNT such as whether or not it has a semiconductor or metal-like electronic structure.

In general, the key feature of nanomaterials is charge confinement which determines their electronic structure. In semiconductor QDs, the carrier confinement is determined by the QD's size. This results in a bandgap and an optical gap that are strongly dependent on the QD diameter. The energy gap increases as the QD size decreases [Klimov (2004)]. The simplest model explaining this behavior is a particle in a spherical box: the smaller the box, the larger the splitting between electronic levels. Similarly, the energy gap of a semiconductor SWCNT is roughly inversely proportional to its diameter. However, this dependence is complicated in nanotubes because tube length and chirality (n,m) also dictate the electronic structure of SWCNTs [Saito *et al.* (1998a)].

When a nano-structure is smaller than the natural length scale of the bulk electron and hole wavefunctions carriers experience strong quantum confinement. In addition to determining the energy gap the strong confinement of QDs and SWCNTs leads to a discreet (quantized) electronic structure, in contrast to the continuous energy bands of bulk materials. The quantized character of the energy spectrum in QDs and SWCNTs makes these systems comparable to atoms and molecules. Yet they are in neither the bulk semiconductor nor the molecular regime. The unique features that originate from their low-dimensionality and strong charge confinement, their quantized and size-tunable energy spectra, place QDs and SWCNTs in between the two traditional types of materials leading to an entirely new range of experimental phenomena.

1.2 Intriguing Phenomena in Quantum Dots and Potential Applications

The 0-D confinement of charge carriers in QDs results in the quantization of their electronic energy levels and is responsible for new physical phenomena. These new physical processes, as well as the tuneability of the optelectronic properties, motivate the use of QDs in burgeoning technological areas such as spintronics [Ouyang and Awschalom (2003)] and quantum computing [Gorman *et al.* (2005); Petta *et al.* (2005)]. QDs have also been incorporated into quantum emitter antennas [Farahani *et al.* (2005)], thermopower [Scheibner *et al.* (2005)] and photovoltaic devices [Nozik (2001); Ellingson *et al.* (2005); Murphy *et al.* (2006); Schaller and Klimov (2004); Schaller *et al.* (2005a, 2006); Klimov (2006)], lasers [Klimov (2006); Klimov *et al.* (2000b)], field-effect transistors [Talapin and Murray (2005)], light-emitting diodes [Coe *et al.* (2002)], and fluorescent biological imaging probes [Dahan *et al.* (2003)]. However, detailed study of these new phenomena is needed before they can be applied further. Sophisticated applications of absorption, fluorescence, Raman, and ultrafast pump-probe spectroscopies are the current mainstream approaches for providing information regarding physical properties and processes in QDs [Nozik (2001); Ellingson *et al.* (2005); Murphy *et al.* (2006); Schaller and Klimov (2004); Schaller *et al.* (2005a, 2006); Klimov (2006); Ellingson *et al.* (2002); Petta *et al.* (2005); Crooker *et al.* (2002); Peterson and Krauss (2006)].

Of particular interest is the recent discovery of multi-exciton states in semiconductor QDs upon absorption of only one high-energy photon as seen in ultrafast pump-probe experiments [Schaller and Klimov (2004); Ellingson *et al.* (2005)]. Predicted [Nozik (2001)] several years prior to their experimental observation [Schaller and Klimov (2004)], the generation of multiple excitons has created intense scientific interest because of its potential for extraordinary improvement in solar-cell efficiency. In traditional photovoltaic materials, photons absorbed at the blue end of the solar spectrum create single high energy excitations; these rapidly relax to the lowest excited state at the red end of the spectrum via interactions with phonons. Thus, significant amounts of solar energy are lost to heat, limiting the maximum thermodynamic efficiency of a traditional solar cell to 33% [Nozik (2001)]. If this energy loss could be used to generate multiple charge carriers, i.e. if a biexciton can be created by absorption of only a single photon, the efficiency could be increased [Nozik (2001)]. In the experiment of Schaller et al. [Schaller *et al.* (2006)], up to seven excitons were observed in PbSe QDs when a single photon was absorbed, showing promise for very high solar cell efficiency.

Carrier multiplication (CM) [Schaller *et al.* (2005a); Califano *et al.* (2004)], also referred to as impact ionization [Nozik (2001)], inverse Auger process [Califano *et al.* (2004); Wang *et al.* (2003)], and multi-exciton generation [Ellingson *et al.* (2005); Murphy *et al.* (2006)], is one of the most remarkable and unique features of semiconductor QDs. Unfortunately, despite intensive study and various proposed mechanisms, clear understanding of the CM process is absent. Moreover, there are contradictory observations of the CM in some types of QDs: some investigations clearly show efficient CM in CdSe nano-crystals [Klimov *et al.* (2007); Schaller *et al.* (2005a, 2006); Klimov (2006)], others do not [Nair and Bawendi (2007)]. Two additional dynamical processes – carrier relaxation through interactions with phonons and Auger recombination – can compete with the CM, diminish its efficiency, and complicate its investigations. In Auger recombination an electron-hole pair can recombine giving up their energy to an electron in the conduction band, increasing its energy [Klimov (2006)]. In contrast to bulk crystals, phonon-mediated carrier relaxation is predicted to be very slow in QDs. Such a slow relaxation can originate from the discrete electronic structure of strongly confined QDs. If there is a mismatch between the energies of electronic transitions and of low-frequency phonons, relaxation is difficult between those electronic states. Consequently it is reasonable to expect that such a mismatch would dramatically slow the charge-carrier relaxation rates [Nozik (2001)] and lead to a so called phonon bottleneck [Harbold *et al.* (2005a); Schaller *et al.* (2005d); Ellingson *et al.* (2005)]. The phonon bottleneck would provide enough time for the CM to take place, favoring the CM processes in QDs with respect to the carrier relaxation.

Electronic absorption spectra of various QDs show distinct peaks indicating well-separated discrete electronic levels, whose splittings are typically at least ten times larger than phonon frequencies [Klimov (2000); Schaller *et al.* (2005c)]. Surprisingly, the reported relaxation times in different QDs [Klimov *et al.* (2000a); Klimov and McBranch (1998); Klimov *et al.* (2000b)] are from sub-picosecond to a few picoseconds. These time-scales are close to the relaxation time-scales in bulk materials in which electronic levels form a continuum. Moreover, pump-probe experiments [Harbold *et al.* (2005a); Schaller *et al.* (2005c)] detect carrier relaxation rates that are higher in smaller QDs which exhibit more dramatic discretization of energy levels. Despite ultrafast relaxation, pump-probe experiments show efficient CM in PbSe and PbS Qds [Schaller *et al.* (2005d); Ellingson *et al.* (2005)]. This result suggests that the phonon bottleneck is not required for efficient CM. The problem of the phonon bottleneck in strongly confined QDs – an intriguing phenomenon predicted theoretically but not observed experimentally – is the main focus of this work and is discussed in detail in the next chapter.

Indeed, carrier relaxation is manifested by excited carrier interactions with phonons. In the adiabatic regime, the data of Raman and photoluminescence (PL) spectroscopies are typically used to evaluate exciton-phonon coupling in QDs [Yoffe (2001)]. For example, the spectra from the 3 nm PbS and PbSe dots show significant broadening originating from coupling to the phonon modes [Kang and Wise (1997)]. The Huang-Rhys factor S, frequently taken as being a rough measure of the strength of the exciton-phonon coupling, has been estimated from the resonant Raman spectrum [Krauss and Wise (1997b)] and from the PL spectra with the fine structure present, including Stokes shifted resonance [Itoh *et al.* (1995)]. The observed values for S, as large as 0.5-0.7, point to strong exciton phonon coupling, which significantly affects carrier dynamics in these QDs [Yoffe (2001)].

The phonon-mediated relaxation dynamics in QDs is the focus of many time-resolved experiments [Nozik (2001); Ellingson *et al.* (2005); Murphy *et al.* (2006); Schaller and Klimov (2004); Schaller *et al.* (2005a, 2006); Klimov (2006); Ellingson *et al.* (2002); Petta *et al.* (2005); Crooker *et al.* (2002); Peterson and Krauss (2006)]. This interest is governed by an important role that carrier relaxation play in a variety of applications. For instance the electron-phonon relaxation rates determine the efficiency of CM, and, consequently, the efficiency of QD-based lasers [Klimov (2006); Klimov *et al.* (2000b)] and photovoltaic devices [Nozik (2001); Ellingson *et al.* (2005); Murphy *et al.* (2006); Schaller and Klimov (2004); Schaller *et al.* (2005a, 2006); Klimov (2006)]. The application of QDs to quantum information processing [Ouyang and Awschalom (2003); Gorman *et al.* (2005); Petta *et al.* (2005); Farahani *et al.* (2005)] is limited by the phonon-induced dephasing of electron [Muljarov *et al.* (2005); Kamisaka *et al.* (2006)] and spin [Petta *et al.* (2005); Semenov and Kim (2004)] excitations. Electron-phonon interactions define linewidths of QD optical spectra [Crooker *et al.* (2002); Peterson and Krauss (2006); Kamisaka *et al.* (2006)]. Inelastic electron-phonon scattering modulates electron tunneling transport through QDs and is responsible for transport blockade and energy loss [Schleser *et al.* (2005)]. Electron-phonon coupling forms the basis of the laser cooling of nanomechanical resonator modes in an embedded QD [Wilson-Rae *et al.* (2004)].

While a variety of time-resolved experimental techniques are used extensively to probe the features of the charge-phonon dynamics in QDs, current theoretical approaches focus on the static characterization of QD structure and spectra [Ellingson *et al.* (2005); Califano *et al.* (2004); Semenov and

Kim (2004); Wilson-Rae *et al.* (2004); Muljarov *et al.* (2005); Efros and Efros (1982); Brus (1984); Puzder *et al.* (2004); Zhou *et al.* (2003)]. To date a little effort has been devoted to the direct modeling of time-resolved experimental dynamics. In this work we report first-principle calculations of electronic structure and phonon-mediated dynamics of carriers in QDs. Applications are made to different materials such as PbSe and CdSe, which are widely used in nanotechnology. We focus specifically on the phonon bottleneck, described above, in strongly confined QDs – a mysterious phenomenon predicted theoretically but not observed experimentally [Nozik (2001); Klimov (2004)].

1.3 Advantages and Challenges of Technological Applications of Carbon Nanotubes

Analogous to QDs, carbon nanotubes are the desirable materials in nanotechnologies. Several aspects of SWCNTs account for their promise in a number of technological applications: a delocalized mobile π-electron system, extraordinary mechanical, electronic, and optical properties, and the ability to tune these properties by specifying the tube geometry. The potential applications include chemical sensors [Kong *et al.* (2000); Snow *et al.* (2005)], mass conveyers [Regan *et al.* (2004)], nanoscale logic gates [Mason *et al.* (2004); Biercuk *et al.* (2004)] and antennas [Dresselhaus (2004)], conductor switches [Chen *et al.* (2002); Terabe *et al.* (2005)], lasers [Set *et al.* (2004a,b)], field-effect transistors, [Misewich *et al.* (2003); Postma *et al.* (2001)], logic circuits [Chen *et al.* (2006)], and drug delivery and cancer therapy [Kam *et al.* (2005)].

Many of the above applications and fundamental studies require manipulation of one particular type of tube with a well-determined diameter and chiral vector. For instance, the bandgap of a semiconducting tube, a critical parameter that needs to be controlled for nanoelectronic applications, is strongly dependent on tube geometry. From an experimental standpoint, however, the specification of individual SWCNTs is difficult because tubes of different species, orientations, and lengths are mixed together in one sample. Tube polydispersion, bundling, poor solubility, and defects complicate the problem even further. Modern catalysts and growth processes allow control of the diameter of synthesized nanotubes to some degree. Samples have been enriched [Chattopadhyay *et al.* (2003)] or separated into pure metallic and semiconductor fractions using dielectrophoresis [Krupke *et al.*

(2003); Chen *et al.* (2003)] or selective reaction chemistry [Krupke *et al.* (2003)]. Separations of individual semiconductor chiralities have been relatively successful using desoxyribonucleic acid (DNA)-wrapping chemistry or reversible chirality-selective SWCNT redox chemistry [O'Connell *et al.* (2005)] which yields enriched samples of only a single large-bandgap chirality [Zheng *et al.* (2003b)]. In spite of these advances, obtaining specific defect-free chiral SWCNT species remains a huge challenge for synthesis and technology.

Just as in the case of QDs, spectroscopic techniques are the current mainstream approach for providing a precise characterization of SWCNT samples. However, a key problem for several years was the lack of fluorescence from nanotube samples, even though, two-thirds of the SWCNTs were predicted to be semiconducting. The major breakthrough occurred in 2002 [Bachilo *et al.* (2002)] when SWCNTs were shown to fluoresce in solution. The authors vigorously sonicated the nanotube bundles in a solution of D_2O and sodium dodecylsulfate, a surfactant. This solution was then centrifuged and the supernatant collected. It was shown that the surfactant created micelles, which encapsulated single SWCNTs. Isolating these SWCNTs removed the nonradiative decay channels provided by the metallic tubes in the sample, and allowed the radiative recombination of the charge carriers. This proved to be a major advancement in the study of SWCNTs, as it was soon followed by sophisticated combinations of absorption, fluorescence, and Raman spectroscopies [Bachilo *et al.* (2002); Weisman and Bachilo (2003); Hartschuh *et al.* (2003a); Sfeir *et al.* (2006)] that make assigning the type of tube routine. The main component of these techniques is spectroscopic identification of the tube diameter and the optical transition energies E_{ii} that specify individual (n,m) nanotubes. Resonant Raman spectroscopy [Dresselhaus and Eklund (2000); Telg *et al.* (2004); Goupalov *et al.* (2006)] delivers an accurate identification of tube diameters and transition energies in an inhomogeneous sample by measuring the radial breathing mode (RBM) frequency. Complementary photoluminescence (PL) experiments provide the (n-m), (2n+m), and (2m+n) family behavior of second absorption peaks and lowest emission peaks, which identify an individual (n,m) tube in a suspension. Yet spectroscopic tube identification is still complicated by inhomogeneous broadening, spectral overlap, and the relatively low efficiency of PL spectra.

Early optical spectra of SWCNTs were interpreted in terms of free electron-hole carriers. However, the electron confinement across a tube leads to high electron-hole binding energies and the formation of strongly-bound excitons. Advanced theoretical studies [Kane and Mele (2004); Spataru *et al.* (2004); Zhao and Mazumdar (2004a); Ferretti *et al.* (2003)] followed by transient spectroscopy and nonlinear absorption experiments [Ma *et al.* (2005b); Korovyanko *et al.* (2004a); Gambetta *et al.* (2006a)] have unambiguously revealed that the photophysics of SWCNTs is dominated by excitons with typical binding energies of $\sim 0.2-0.5$ eV [Wang *et al.* (2005a); Rafailov *et al.* (2005)], about a hundred times larger than that in bulk semiconductors and nearly ten times larger than in QDs [Yoffe (2001)].

The low PL efficiency of SWCNTs is another challenge tightly connected with the excitonic structure in nanotubes. The 1-D charge confinement of SWCNTs leads to sub-band quantization of electronic states and a series of energy manifolds. Each manifold contains both optically allowed (bright) and optically inactive (dark) exciton states, as well as a continuum band [Zhao *et al.* (2006); Wang *et al.* (2006b)]. The number and energy position of dark excitons with respect to the bright strongly affect the photoluminescence quantum yield. Recent theories [Zhao and Mazumdar (2004a); Perebeinos *et al.* (2005c)] and magneto-optic experiments [Shaver *et al.* (2007)] explain the typically low yield of photoluminescence in SWCNTs by pointing to the existence of dark excitons, below the first optically bright exciton state, trapping most of the exciton population. However, existing optical experiments have focused on the lowest transitions E_{ii} (E_{11}, E_{22}, and so on) from the excitonic band edges. Study of the entire excitonic spectra of SWCNTs still remains incomplete.

In addition to probing excitonic phenomena recent research has shown the importance of electron-phonon coupling (vibrational effects) in carbon nanotubes [LeRoy *et al.* (2004b); Plentz *et al.* (2005a); Htoon *et al.* (2004); Fantini *et al.* (2004b); Gambetta *et al.* (2006a); Perebeinos *et al.* (2005b); Plentz *et al.* (2005a); Habenicht *et al.* (2006)]. Similar to the case of QDs, there is a substantial electron-phonon interaction in SWCNTs. This follows from several experiments: strong phonon-assisted bands identified in PL spectra [Plentz *et al.* (2005a); Htoon *et al.* (2004)] and coherent phonon generation [Gambetta *et al.* (2006a)] detected in time-resolved spectra. It is important to note that an interplay between strong Coulomb correlations and electron-phonon coupling is a typical feature in all 1-D systems including conjugated organic [Heeger *et al.* (1988); Tretiak and Mukamel (2002)] and organometallic [Batista and Martin (2005)] polymers, mixed-valence chains [Dexheimer *et al.* (2000)], and organic molecular materials [Klessinger and Michl (1995); Balzani and Scandola (1991); Herzberg

(1950)] in general. The relationship between excitonic and vibrational effects frequently gives rise to spontaneous symmetry breaking (Peierls distortion). This leads to rich physical phenomena such as metal-semiconductor transitions, solitons, polarons, and breathers, thereby enhancing technological applications. The exploration of these phenomena in the context of SWCNTs is complicated by sample quality issues and the significant numerical effort involved in quantum-chemical studies. Many questions still remain.

Excitonic and vibronic effects are thus equally important for a correct description of photoinduced dynamics in nanotubes. Both of these phenomena need to be clearly understood for the efficient functionality of future SWCNT-based electronic devices to be achieved. Most theoretical investigations of SWCNTs have been carried out using model calculations or empirical tight-binding approximations [Saito *et al.* (1998a); Zhao and Mazumdar (2004a); Perebeinos *et al.* (2004); Chang *et al.* (2004)]. A few first-principle studies focused on snapshots of the electronic structure of narrow-diameter SWCNTs have recently been published. Aside from one limited effort [Perebeinos *et al.* (2005b)], modeling of excited-state molecular dynamics has not been pursued because of the significant numerical expense. has yet to be realized.

This work presents extensive quantum-chemical studies of excitonic and vibrational effects in SWCNTs, including free carrier relaxation and dephasing. We use finite-size molecular-type approaches to study excitonic effects, as well as bulk-like approaches based on periodically boundary conditions to simulate relaxation and dephasing times. We specifically focus on the dependence of exciton-vibrational phenomena on tube chiralities and diameters to reveal the intricate details of excited state phenomena in SWCNTs which can be compared with experimental results.

1.4 Excitonic Character and Numerical Approaches to QDs and SWCNTs

As discussed in the previous sections, carrier-phonon interaction is substantial in QDs and SWCNTs and strongly affects carrier dynamics in both nanomaterials. Because of the complexity of the considered systems and simulated phenomena, we focus on the dominant features accurately while other processes are neglected in order to reduce numerical expenses.

With QDs, our main focus is on the explanation of the rapid carrier relaxation and the absence of the phonon bottleneck. For this purpose, *nonadiabatic* dynamics is used to simulate carrier hops between different electronic levels, describing the phonon-induced relaxation process. It also allows for the calculation of relaxation rates that can be compared with experiment. Analogous simulations of nonadiabatic dynamics are used to calculate phonon-induced intraband relaxation rates of free carriers and dephasing times in SWCNTs. However, with SWCNTs we also explore coupled exciton-vibrational dynamics and focus on the question of how the geometrical changes (bond-lengths and bond angles) induced by photoexcitation affect the electronic structure in SWCNTs of different size. In this context we are interested in describing the excitonic structure, in particular exciton coherence size and exciton localization on the tube, as a function of SWCNT diameter and chirality. We also evaluate the carbon nanotube exciton-phonon coupling and its size-scaling which can be compared with experimental data. This information can be extracted from *adiabatic* dynamics to represent the relaxation of an exciton without switching electronic levels. In this respect, the book is organized as follows. Chapters 1.4 and ?? are dedicated to simulations of nonadiabatic dynamics in CdSe and PbSe QDs and in the smallest available SWCNTs (6,4) and (8,4), respectively. Chapter 3.3 represents results of adiabatic exciton-vibrational dynamics in fifteen different SWCNTs.

In addition to carrier-phonon interactions, charge confinement in both nanostructures leads to strong Coulomb interaction between electrons and holes, and, therefore, to excitonic effects. However, the character of excitons and their computational treatment are different in the two systems. For example, in strongly confined QDs, both the Coulombic and the kinetic energies of electrons and holes rise with decreasing QD radius (R). This combined effect is due to the confinement of the charge motion in all three spatial directions. In order to make the comparison of the kinetic and potential energies we consider them as function of the ration R/a_{B}. Here the constant a_{B} is Bohr radius of electron-hole pair for a specific material:

$$a_B = \frac{\epsilon \hbar^2}{\mu e^2},$$

where $\mu^{-1} = m_e^{*-1} + m_h^{*-1}$ is the electron-hole reduced mass, ϵ stands for dielectric constant of the specific material, e electron charge. In these terms, kinetic energy is proportional to $(a_{\mathrm{B}}/R)^2$ while Coulombic energy is proportional to a_{B}/R [Klimov (2004)]. Consequently, the Coulomb energy is much smaller than the kinetic energy for small QDs and can be treated as a first-order perturbation [Ekimov *et al.* (1993, 1985); Efros and Rosen (2000); Brus (1984)]. Therefore, to a zeroth-order approximation, carrier-phonon dynamics in strongly confined QDs can be considered as independent, phonon-mediated electron and hole dynamics. Theoretical models based on this approximation are reviewed in Section 2.1.

This approximation allows us to apply density functional theory (DFT) to calculate non-adiabatic electron-phonon coupling and to simulate real-time atomistic relaxation dynamics of electrons and holes in relatively large semiconductor QDs (up to 222 atoms) in 4 ps time-domain. While the sizes of the systems we study are close in size to QDs typically investigated experimentally, our numerical calculations have an accuracy similar to first-principle quantum mechanical methods. These simulations are made possible by the recent implementation [Craig *et al.* (2005)] of trajectory surface hopping (TSH) [Tully (1990); Hammes-Schiffer and Tully (1994); Parahdekar and Tully (2005)] within the time-dependent Kohn-Sham (TDKS) theory [Marques and Gross (2004); Baer and Neuhauser (2005)]. The approach was developed in our research group at the University of Washington under the leadership of Oleg Prezhdo with contributions from other group members (Colleen Craig, Walter Duncan, and Brad Habenicht). Section 2.2 explains the method in detail.

Previously, the TSH methodology was successfully used for simulations of the electron back-transfer process in the alizarin-TiO2 system [Duncan and Prezhdo (2007); Craig *et al.* (2005)] which is a model of the charge-separation component in Grätzel-type solar cells [Oregan *et al.* (1991)]. It was also applied to investigations of photoinduced electron and hole dynamics in (7,0) SWCNT where it showed agreement with experimental timescales and provided a wealth of information about electron-phonon coupling in SWCNTs [Habenicht *et al.* (2006)]. Studying dephasing in QDs, we based our approach on TDKS *ab initio* molecular dynamics within the framework of the optical response function and semiclassical formalisms. When we investigated vibrationally induced dephasing times in PbSe QDs, we found them to agree perfectly with the experimentally measured homogeneous line width of optical transitions [Kamisaka *et al.* (2006)]. Recently, we extended this study to calculate coherence loss dependence on temperature and pressure in CdSe and PbSe QDs [Kamisaka *et al.* (2008)]. Finally, we successfully applied the TSH methodology to study phonon-mediated electron and hole relaxation dynamics in a small (1 nm in diameter) PbSe QD and explain the sub-picosecond time scale of the relaxation and the absence of the phonon bottleneck [Kilina *et al.* (2007)].

The current work presents an application of fully atomistic real time investigations to carrier relaxation in QDs and generalizes them for larger dots (up to $\sim$ 2 nm in diameter) made of different materials such as PbSe and CdSe which are frequently used in experiments. Section 2.3 contains the calculated results which were partially published in Ref. [Kilina *et al.* (2007); Kamisaka *et al.* (2006)] and in review for publication in the Journal of Physical Chemistry C at the time of this manuscript. Subsections 2.3.1 and 2.3.2 present an overview of the electronic and structural properties ofthe host materials of PbSe and CdSe QDs. Densities of states at zero Kelvin temperature and ambient temperatures for both QDs are also presented in these subsections. We simulate the optical spectra of QDs and compare our results with available experimental data. An intricate detail – the S-P transition in the spectra of PbSe QDs, which has been probed experimentally but not yet explained – is analyzed and discussed in Subsection 2.3.3. In QDs made of both materials, we observe relatively high electron and hole state densities that are attributable to surface reconstructions. While many of these states show only weak optical activity that results in discrete peaks in the optical spectra, most facilitate the electron-vibrational relaxation assisted by strong non-adiabatic electron-phonon coupling. The calculated properties of phonon modes and the associated electron and hole relaxation rates in CdSe and Pbse QDs are detailed in Subsections 2.3.4 and 2.3.5, respectively. The fine features of the carrier relaxation dynamics mechanism and comparison of dynamics in CdSe and PbSe QDs are extensively discussed in Subsection 2.3.6. Our simulations generate valuable insights into QD properties and reconcile the seemingly contradictory observations of wide optical line spacing and the lack of the phonon bottleneck to relaxation. Section 2.4 summarizes our main simulation results.

In contrast to QDs electron-hole pairs in SWCNTS are tightly bound excitons whose interaction energy makes a significant contribution to the Hamiltonian of the systems. Therefore, excitonic effects can neither be neglected nor treated as a small perturbation to the bandgap as in the case of QDs. Section 4.1 briefly discusses excitonic properties and the photo-physics of SWCNTs from the experimental point of view. An accurate description of photoinduced dynamics in SWCNTs requires numerical methods to include both strong Coulombic and exciton-phonon couplings.

The approach used for QDs demonstrates that TDKS (DFT-based molecular dynamics) can be utilized to simulate electron- and hole-phonon dynamics where electrons and holes are independent carriers [Kilina *et al.*

(2007); Habenicht *et al.* (2006)]. However, DFT does not provide accurate Coulomb energies due to the incomplete elimination of the self-interaction by pure DFT functionals [tri (1990); Gross *et al.* (1996)]. Long-range nonlocal and nonadiabatic density functional corrections (such as hybrids including a portion of the exact Hartree-Fock exchange) in combination with time dependent density functional theory (TDDFT) – based on the linear response approximation – are routinely used to remedy the problem [Champagne *et al.* (2006); Tretiak *et al.* (2005)]. Note that TDDFT differs from TDKS: the first method has corrections to the exchange and Coulomb energies without considering dynamics; the second incorporates molecular dynamics while failing in the proper description of electron-hole interactions. Although TDDFT provides accurate electron-hole correlations, it is an extremely numerically expensive method that can hardly be used to describe excited state dynamics in large systems such as the SWCNTs studied here. Nonetheless, snapshot calculations of excitonic features in nanotubes are possible and reported in Ref. [Tretiak (2007a)]. A short review of available methods that take into account nanotube excitonic effects, vibrational effects, or both, is presented in Subsection 4.1.2.

Here we use the excited-state molecular dynamics (ESMD) technique [Tretiak *et al.* (2002); Tretiak and Mukamel (2002)] for the computation of the excited state potential surfaces and for simulations of photoexcitation dynamics in SWCNTs. This methodology was recently developed by Sergei Tretiak at Los Alamos National Laboratory and successfully applied to many conjugated molecular materials [Mukamel *et al.* (1997); Tretiak *et al.* (2000b); Franco and Tretiak (2004); Tretiak *et al.* (2003); Wu *et al.* (2006)]. Having similar machinery as the TDDFT, this technique combines semi-empirical Hamiltonians [Dewar *et al.* (1985); Zerner (1996); Baker and Zerner (1991)] with time-dependent Hartree-Fock (TDHF) formalism [Stratmann *et al.* (1998); Mukamel *et al.* (1997)]. The TDHF implements essential electronic correlations and excitonic effects with accuracy and only moderate numerical expense. Utilization of the analytical gradients of the excited-state potential energy surfaces [Furche and Ahlrichs (2002)] in the ESMD package allows investigation of vibrational phenomena as well as excited state optimizations and dynamics [Tretiak *et al.* (2002, 2003)]. Section 4.2 describes the ESMD method in detail.

This work presents the results of ESMD calculations that address excited state phenomena in SWCNTs based on several original publications. These include the quantification of exciton-phonon coupling and Huang-Rhys factors [Shreve *et al.* (2007a)], the characterization of delocalized non-excitonic transitions (E_{33} and E_{44}) [Araujo *et al.* (2007)], the effects of Peierls distortion and exciton self-trapping [Tretiak *et al.* (2007b); Kilina and Tretiak (2007)], and the delocalization properties of cross-polarized excitons (E_{12} and E_{21}) [Kilina *et al.* (2008)]. Section 4.3 summarizes the application of the ESMD approach to several excitonic bands, focusing on bright photo-induced excitons and their coupling to different vibrational modes of SWCNTs. We model in detail the excited-state properties in ten chiral and five zigzag carbon nanotubes. The structures of the tubes we study are described in Subsection 4.2.4. Calculated electronic spectra of SWNTs and their dependence on tube diameter (curvature) and chirality are described in subsection 4.3.1. Spatial structure of excitons and their delocalization properties are presented in Subsection 4.3.3. Subsection 4.3.4 analyzes lattice relaxation upon photoexcitation: a Peierls distortion and its effect on excitonic structure resulting in exciton self-trapping. The numerical estimation of exciton-phonon coupling and Huang-Rhys factors in different SWCNTs, in addition to their comparison with experimental data, are presented in Subsection 4.3.5. Section 4.4 summarizes our results.

Chapter 2

Electronic Structure and Phonon-Induced Carrier Relaxation in CdSe and PbSe Quantum Dots

Phonon-induced relaxation of a charge to its ground state and other dynamics of excited charge carriers affect many important characteristics of nanoscale devices such as switching speed, luminescence efficiency, and carrier mobility and concentration. Therefore, a better understanding of the processes that govern such dynamics has important and fundamental technological implications. These phenomena are investigated experimentally through intraband transitions induced by carrier-phonon interactions in QDs and probed effectively by various time-resolved spectroscopy methods. Yet fundamental understanding of the underlying physics responsible for carrier dynamics and the specific role that phonons play in the relaxation mechanisms in QDs is still lacking. For example, it remains a mystery why the expected electron-phonon relaxation bottleneck is not experimentally observed and what effect the QD host material has on carrier relaxation, carrier multiplication, and Auger processes. Answering these questions can be facilitated significantly by theoretical and computational studies which provide a description of a system on the atomic level that is difficult – if not impossible – to probe experimentally.

We begin this chapter with a discussion of the basic theoretical concepts necessary to understand electronic structure and electron-phonon coupling in nanocrystals followed by an overview of experimental reports of carrier relaxation rates and the phonon bottleneck phenomena in CdSe and PbSe QDs. We then present our numerical approach and a detailed analysis of the simulation results of carrier-relaxation in these two types of QDs. The majority of these results were originally published in Ref. [Kilina *et al.* (2007); Kamisaka *et al.* (2006)] or in review for publication in the Journal of Physical Chemistry C at the time of this manuscript. The conclusions section summarizes the key results within a broader perspective of photoinduced excitation dynamics in QDs.

2.1 Carrier Dynamics in Quantum Dots: Review of Exciting Theoretical and Experimental Approaches

2.1.1 *Theoretical Considerations of Electronic and Optical Properties of Quantum Dots*

The earliest and simplest treatment of the electronic states of a QD is based on the effective mass approximation (EMA) [Efros and Efros (1982)]. The EMA rests on the assumption that if the QD is larger than the lattice constants of the crystal structure, then it will retain the lattice properties of the infinite crystal and the same values of the carrier effective masses. Thus, the electronic properties of the QD can be determined by simply considering the modification of the energy of the charge carriers produced by the quantum confinement.

According to Bloch's theorem, the electronic wavefunction in bulk crystal can be written as

$$\Psi_{nk}(\mathbf{r}) = u_{nk}(\mathbf{r})exp(i\mathbf{k}\cdot\mathbf{r}), \tag{2.1}$$

where u_{nk} is a function with the periodicity of the crystal lattice. Here the wavefunctions are labeled by the band index n and wave vector k. The energy of these wavefunctions is typically described in a band diagram, a plot of E versus k. In EMA, the bands are assumed to have simplified parabolic forms near extrema in the band diagram. For a direct gap semiconductor such as CdSe both the valence band (VB) maximum and conduction band (CB) minimum occur at $k = 0$ (See Fig. 2.2). In this approximation, the carriers behave as free particles with an effective mass m_e^* for electrons and m_h^* for holes. Graphically, the effective mass accounts for the curvature of electronic levels with respect to k. Physically, the effective mass attempts to incorporate the complicated periodic potential experienced by the carriers in the lattice. Thus, EMA treats electrons and holes in a crystal as free particles but with different masses.

If the QD diameter is much larger than the lattice constant of the host material, the single-particle wavefunction of QD can also be represented as a linear combination of Bloch functions

$$\Psi_{sp}(\mathbf{r}) = \sum_k C_{nk}u_{nk}(\mathbf{r})exp(i\mathbf{k}\cdot\mathbf{r}), \tag{2.2}$$

where C_{nk} are expansion coefficients that ensure the sum satisfies the spherical boundary conditions of the QD. Furthermore, if we assume that the functions u_{nk} have a weak k dependence because of the finite QD size, then Eq. 2.2 can be rewritten as

$$\Psi_{sp}(\mathbf{r}) = u_{n0}(\mathbf{r}) \sum_{k} C_{nk} exp(i\mathbf{k} \cdot \mathbf{r}) = u_{n0}(\mathbf{r}) f_{sp}(\mathbf{r}), \tag{2.3}$$

where $f_{sp}(\mathbf{r})$ is the single-particle envelope function. This approximation is sometimes called the envelope function approximation [Klimov (2004)] after $f_{sp}(\mathbf{r})$. Thus, within EMA, the QD electron and hole wavefunctions can be represented as a product of periodic Bloch functions and envelope wavefunctions. The Bloch function describes carrier motion in the periodic potential of a crystalline lattice, whereas the envelope function describes the motion in the QD confinement potential.

Because the periodic function u_{n0} can be determined within the tight-binding approximation as a sum of atomic wavefunctions $\varphi_{n,k}$ (summed over lattice sites), the QD problem is reduced to determining the envelope functions f_{sp}; thus, the electronic properties are determined by solving the Schrödinger equation for a particle in a 3-D box. The zeroth-order approximation is a perfectly spherical QD with infinite potential walls at the surface. In this case, the envelope functions of the carriers in a QD are given by the well-known particle-in-a-sphere solution where the masses of electrons and holes are equal to their effective masses in bulk crystals. Due to the symmetry of the problem, each electron and hole wavefunction can be described by an atomic-like orbital labeled by the quantum numbers: n (1, 2, 3...), l (S, P, D...), and m. The first quantum number n represents the number of the state within a series of states of the same symmetry; l corresponds to the orbital momentum of the envelop function; and m denotes the projection of the total angular momentum along a magnetic axis.

There is also a Coulomb interaction between an electron and a hole that leads to excitons in QDs. However, in the case of strong confinement, when the QD size is small compared to the de Broglie wavelength of electrons in bulk crystals, the confinement energy of each carrier scales as $(a_{\mathrm{B}}/R)^2$ (the energy of a particle in a spherical box) while the Coulomb interaction between an electron and a hole scales as a_{B}/R, where a_{B} is exciton Bohr radius. Consequently, the quadratic confinement term dominates, and the exciton states are well-described in terms of one-particle electronic states $\Psi_e(\mathbf{r_e})$ and hole states $\Psi_h(\mathbf{r_h})$. Thus in zeroth-order approximation, the electron-hole pair states in QDs can be written as a product:

$$\Psi(\mathbf{r}_e, \mathbf{r}_h) = \Psi_e(\mathbf{r_e})\Psi_h(\mathbf{r_h}) \tag{2.4}$$

Using Eq. 2.3 one gets

$$u_{n0}(\mathbf{r}) f_{sp}(\mathbf{r}) = C\left(u_c \frac{j_{L_e}(k_{n_e,L_e} r_e) Y_{L_e}^{m_e}}{r_e}\right)\left(u_c \frac{j_{L_h}(k_{n_h,L_h} r_h) Y_{L_h}^{m_h}}{r_h}\right) \tag{2.5}$$

with energies

$$E(n_e, L_e, n_h, L_h) = E_g + \frac{\hbar^2}{2R^2}\left\{\frac{\alpha_{n_e,L_e}^2}{m_e^*} + \frac{\alpha_{n_h,L_h}^2}{m_h^*}\right\}, \tag{2.6}$$

where $\alpha_{n,L}$ are roots of L-th order spherical Bessel functions $j_L(k_{n,L}r)$.

In cases where the effective masses of the electron and hole are similar as in PbSe QDs, both CB and VB states are usually characterized by two quantum numbers: n, the principal quantum number (1, 2, 3, etc.), and L, the orbital angular momentum (S, P, D, etc.). Thus, electron and hole states are labeled as nL_e ($1S_e$, $2S_e$, $1P_e$, etc.) and nL_h ($1S_h$, $2S_h$, $1P_h$, etc.) since other quantum numbers are identical for electron and hole states involved in optical transitions . In contrast, when the effective masses of electrons and holes are different as in CdSe QDs, a description of heavier carrier states requires the use of an additional quantum number F which is the total carrier angular momentum $F = L + J$ (i.e., the sum of the Bloch function (J) and envelope function (L) total momenta). Here $J = l + s$ where s is the spin and l is the orbital momenta of atomic orbitals $\varphi_{n,k}$. The projection of F along a magnetic axis is represented by m_F and ranges from $-F$ to F. Taking CdSe as an example, a commonly used notation [Klimov (2004)] employs nL_e for the electron states and nL_F for the hole states; electron states are represented by $1S_e$, $2S_e$, $1P_e$, etc., and hole states by $1S_{1/2}$, $1S_{3/2}$, $1P_{1/2}$, etc. For optical transitions in ideal spherical QDs, the selection rules are $\Delta n = 0$, $\Delta L = 0; \pm 2$, and $\Delta m = 0; \pm 1$. These ideal selection rules can be broken by nonspherical QDs and strong interference between states. For visualization, Fig. 2.1 shows the experimental absorption spectra of CdSe and PbSe QDs and the notation of the most pronounced optical transitions in accordance to EMA.

Of course, the simple EMA model gives only a qualitative description of electronic structure of QDs. For more quantative calculations of QD optical spectra, various approaches based on EMA have taken into account the Coulomb interaction between electrons and holes as a first-order correction. These include perturbation theory [Chestnoy *et al.* (1986); Brus

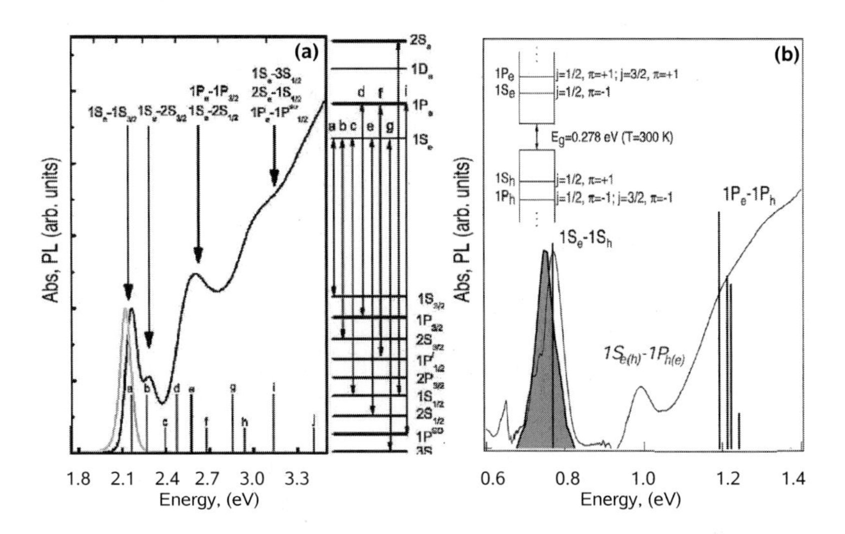

Fig. 2.1 Experimental absorption and emission spectra of CdSe and PbSe quantum dots from Refs [Sewall *et al.* (2006); Harbold *et al.* (2005a)]. Peaks are interpreted and labeled in accordance to the effective mass calculations. **(a)** Absorption spectrum of CdSe quantum dots (black line) and photoluminescence spectrum (grey line) adopted from Ref. [Sewall *et al.* (2006)]. The tall or short sticks in the spectrum reflect high or low oscillator strength transitions while the level diagram depicts the transitions into specific excitonic states. **(b)** The absorption (solid line) and photoluminescence (shaded region) spectra for 6 nm PbSe QDs in the colloidal silica film, adopted from Ref. [Harbold *et al.* (2005a)]

(1984)], variational calculations [Ekimov *et al.* (1985); Takagahara (1993)], matrix diagonalization [Hu *et al.* (1990)], and Monte Carlo methods [Pollock and Koch (1991)]. Furthermore, the simple EMA model has been significantly improved by including the effects of band nonparabolicity, hole-state mixing [Ekimov *et al.* (1993); Efros and Rosen (2000)], finite barrier heights [Kayanuma and Momiji (1990)], and surface polarization [Banyai *et al.* (1992)]. Incorporation of these effects to EMA is done by the $\mathbf{k} \cdot \mathbf{p}$ approach that has been commonly used to calculate the electronic structure of bulk semiconductors. In the effective mass $\mathbf{k} \cdot \mathbf{p}$ approach, the QD wavefunctions are expanded as a linear combination of bulk Bloch functions at a particular point in k-space, typically $k = 0$ [Sercel and Vahala (1990); Banin *et al.* (1998)]. The $\mathbf{k} \cdot \mathbf{p}$ approach has been used extensively for QDs to model optical spectra and the dependence of optical energy-gap on QD size.This approach is in agreement with experimental data of QDs larger than 2 nm in diameter [Klimov (2004); Yoffe (2001)]. The validity of the $\mathbf{k} \cdot \mathbf{p}$ approach has been questioned [Zunger (2001); Fu *et al.* (1998)], but it has also been vigorously defended [Efros and Rosen (1997, 2000)]; the debate continues.

An alternative theory based on direct diagonalization and pseudopotentials has been proposed as a better theoretical approach to calculate the electronic states in QDs [Zunger (2001); Fu *et al.* (1998)]. This method involves three steps. First, one-electron wavefunctions and energies are determined by solving the Schrödinger equation in a plane-wave basis set with explicit spin-orbit coupling and local potential $V(\mathbf{r})$, represented by the sum of screened atomic pseudopotentials fitted to reproduce bulk properties [Wang and Zunger (1995)]. Coulomb and exchange integrals between pairs of single-particle states from the CB and VB are calculated to introduce excitonic effects in QDs. To calculate these matrix elements, a dielectric function ε is chosen to represent the bulk-like environment inside the QD and is parametrized specifically in accordance with the surface to simulate the dielectric function of the surrounding material. Third, a configuration interaction (CI) Hamiltonian is constructed in the basis of Slater determinants and composed of asymmetrical products of N electron and N hole single-particle wavefunctions. Solving the eigenvalue problem for this CI Hamiltonian provides accurate electronic structure and optical transitions in Qds. It is important to note that in this method Coulomb interaction in QDs is treated non-perturbatively within the many-body CI calculation. Results for PbSe QDs of size $1.5 - 3$ nm show a relatively small value of Coloumb correction, below $5 - 10\%$ for many-body energy gap versus single-particle energy gap (e.g. $E_{coulomb} = 39$ meV vs $E_g = 855.5$ meV) [An *et al.* (2006); Franceschetti *et al.* (2006)]. Experimentally measured exciton binding energies in different QDs are also in the range $20 - 50$ meV [Yoffe (2001)]. This supports the EMA assumption that the Coulomb term is much smaller than the kinetic term in QDs.

The atomic pseudopotential approach has been applied to zincblende [An *et al.* (2006)], wurtzite [Wang *et al.* (2003)], and diamond-like [Lany *et al.* (2007)] semiconductor QDs. It takes into account inter-band coupling, spin-orbit coupling, and dielectric screening, and has therefore been used for the calculation of Auger splitting and biexciton binding energies in QDs [Califano *et al.* (2004); Franceschetti *et al.* (2006); Califano *et al.* (2007)]. This method predicts the density of states, optical spectra, and bright and dark excitons; it accounts for the high multiplicity of band-eigenstates and can be used to describe charged QDs. Therefore, the method of atomic pseudopotentials suggests a more subtle structure of excited states compared to EMA and $\mathbf{k} \cdot \mathbf{p}$ theories. However, this method does not take into account phonon-effects and cannot be used for simulating carrier dynamics in QDs. Also, the crucial effects of QD surface reconstruction and passivation are not explicitly included in this model.

The surface effects of QDs could be calculated by first-principle methods such as DFT. Unfortunately, DFT is a very numerically expensive method and its application is limited for large inorganic systems. So far there have been only a few DFT-based calculations of the electronic structure of QDs [Puzder *et al.* (2004); Kilina *et al.* (2007); Kamisaka *et al.* (2006)]. The first DFT-based investigations of CdSe clusters [Puzder *et al.* (2004)] have shown that geometry optimization leads to strong surface reconstructions that significantly deviate from the bulk geometry although the bulk-like structure is preserved inside the QD. Such a strong surface reconstruction compensates for dangling bonds on the unpassivated surface of the simulated QD and opens the energy gap in QDs [Kilina *et al.* (2007)]. The details of QD electronic structure obtained by DFT method with hybrid functionals are discussed in Subsection 2.1.2. Ligands are typically used in experimental QD samples to suppress dangling bonds on the QD surface, but their effect on electronic structure has not been rigorously studied yet.

2.1.2 *Dependence of Electronic Structure on a Host Material: CdSe and PbSe Quantum Dots*

Here we consider CdSe and PbSe QDs, two intensively studied nanostructures that have the potential to act as the basic building blocks of electronic and optical devices of the future. CdSe QDs are among the most exhaustively investigated semiconductor QDs due to their high luminescence efficiency, size-tunable emission in a wide range of visible light, and reasonably long excited-state lifetimes [Klimov (2004); Klimov and McBranch (1998); Klimov *et al.* (2000b, 1995); Nozik (2001); Yu *et al.* (2004); Shimizu *et al.* (2001); Kuno *et al.* (2001); Guyot-Sionnest *et al.* (2005); Mittleman *et al.* (1994); Ekimov *et al.* (1993); Prabhu *et al.* (1995); Ispasiou *et al.* (2001)]. Recent observations of efficient CM in lead chalcogenide (PbS, PbSe, and PbTe) QDs [Schaller and Klimov (2004); Ellingson *et al.* (2005); Murphy *et al.* (2006)] have generated strong interest in these nanocrystals, especially in PbSe QDs. In addition, PbSe QDs are among the few materials that can provide size-tunable optical transitions at near-infrared (NIR) wavelengths, important for technological [Talapin and Murray (2005); Wise (2000)] and biological applications [Bakueva *et al.* (2004)]. The CM process has also been reported in CdSe QDs [Schaller *et al.* (2005c); Klimov *et al.* (2000b); Schaller *et al.* (2005b)], but other investigators doubt the existence of CM in this type of nanocrystal [Nair and Bawendi (2007)]. These varied and somewhat contradictory investigations raise questions about how CM and other related processes such as Auger recombination and carrier relaxation depend on the QD host material and its specific electronic properties.

In bulk crystals, geometrical and electronic structure of these two materials are very different. PbSe and CdSe bulk crystals normally exhibit colloidal QD rocksalt [Wise (2000)] and wurzite [Murray *et al.* (1993)] lattice symmetry. These differences in crystal lattice are reflected in drastically different electronic band structures. For instance, CdSe is a direct-gap semiconductor with a fundamental bandgap at the center of the Brillouin zone (Γ-point, $k = 0$). Its bandgap energy is ~ 1.75 eV which is at the red range of visible spectra. In contrast, PbSe bulk crystal has a narrow bandgap of $0.17 - 0.4$ eV (NIR) at the L point of the Brillouin zone [Wise (2000)]. According to EMA, the electronic structure of QDs made from PbSe and CdSe materials must be different as well.

In CdSe QDs, it is supposed that the effective mass of holes is nearly three times larger than that of electrons. This factor originates from different degeneracies in the orbital quantum number for electron and hole states. The highest state at the CB of CdSe originates from Cd 5s-atomic orbitals and is only twofold degenerate at $k = 0$, including spin. In contrast, the lowest state at the VB arises from Se 4p-atomic orbitals and is sixfold degenerate at $k = 0$. However, due to the strong spin-orbit coupling (0.42 eV) in CdSe, its VB degeneracy at $k = 0$ is split into $p_{3/2}$ and $p_{1/2}$ sub-bands, where the index refers to the total angular momentum $J = l + s$ ($l = 1, s = 1/2$), with l and s the orbital and spin momenta. Away from $k = 0$, the $p_{3/2}$ band is further split into $J_m = \pm 3/2$ (called heavy holes) and $J_m = \pm 1/2$ (called light holes) sub-bands, where J_m is the projection of J. In addition, the crystal field splitting which occurs in materials with a wurtzite lattice causes the degeneracy at $k = 0$ for the $J = 3/2$ sub-band to be lifted by ~ 25 meV. The generic form of the energy levels of CbSe QDs is shown in Fig. 2.2. In reality, CdSe electronic spectra are complicated further by inter-band mixing [Klimov (2004)]. Thus, the VB and CB in CdSe are asymmetric with respect to each other due to closely spaced hole levels and sparser electron levels. According to the EMA, CdSe QDs inherit this band asymmetry from the bulk crystals.

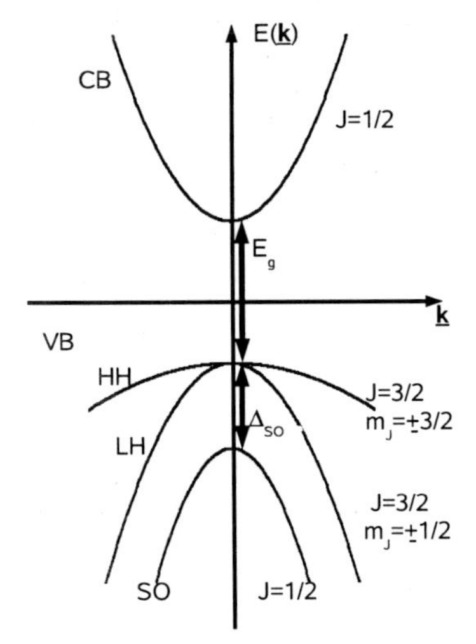

Fig. 2.2 Schematic plot of band diagram for diamond-like semiconductor within the effective mass approximation. The total momentum J is indicated for each band. The bandgap is labeled as E_g. The conduction band is labeled as CB. The valence band is represented by three sub-bands, labeled as HH, LH, and SO. Heavy holes (HH) and light holes (LH) sub-band maxima are separated from the SO sub-band maxima by spin-orbit splitting Δ_{SO}.

In contrast, the bulk PbSe has both its VB maximum and its CB minimum located at the fourfold-degenerate L-point in the Brillouin zone. Therefore, the effective masses of electrons and holes are nearly equal in PbSe QDs, and the VB and CB have a mirror-like symmetry in accordance with the $\mathbf{k} \cdot \mathbf{p}$ calculations [Kang and Wise (1997)] of the PbSe-bulk band structure. However, recent results of an atomistic pseudopotential method [An *et al.* (2006)] raise doubts about the symmetry between the VB and CB in PbSe nanocrystals. Contrary to $\mathbf{k} \cdot \mathbf{p}$ calculations, this approach shows a lack of symmetry between electron and hole states near the energy gap: the VB of PbSe QDs has a higher density of states than the CB. This calculation is supported by the presence of three VB maxima and just one CB minimum in the spectra of bulk PbSe. However, this asymmetry is pronounced only for states near the bandgap; with a photon energy of few bandgaps this method demonstrates symmetry between electron and hole states. Clearly, the electronic structure of PbSe QDs is still under debate.

It is also important to note that the effective masses of electrons and holes in PbSe are smaller than those in CdSe QDs. Table 2.1 compares the exciton Bohr radius a_B (maximal extension of electron-hole pair wavefunction) in CdSe and PbSe bulk materials adopted from Ref. [Wise (2000)]. Thus, given the same nanocrystal radius (R), it is easier to reach a strong

R/a_B	CdSe	PbSe
1	30%	5%
0.3	90%	15%
0.1		45%
a_B (nm)	6	46

confinement regime ($R < a_B$) and, consequently, more discrete and simpler optical spectra in PbSe, rather than in CdSe QDs (see Fig. 2.1). Experimentally, it is possible to achieve $R/a_B \approx 0.04$ or even $R/a_B \approx 0.02$ in PbSe QDs, while the value of $R/a_B \approx 0.16$ is minimal in commonly studied CdSe QDs [Wise (2000)]. As a result, the degree of confinement possible with the PbSe QDs can be many times stronger than in CdSe nanocrystal materials. As one measure of this, the confinement energy in PbSe QD can easily be several times its bulk bandgap, compared to about one-half in the case of CdSe. However, the potential challenge for studies of nanocrystal physics is how to achieve strong quantum confinement of charge carriers without the properties being dominated by the surface of the QD. Because the influence of the surface on QD properties is complicated and remains controversial, reducing this influence could be valuable in determining intrinsic QD properties. For this reason PbSe QDs have an advantage over CdSe Qds; the surface effects are much less pronounced because PbSe QDs crystallize in the cubic sodium chloride structure with every atom at a site of inversion symmetry. The hexagonal CdSe lattice does not exhibit inversion symmetry. This is illustrated in Table 2.1 adopted from Ref. [Wise (2000)]

The unique and different features of these two nanocrystals, in particular the supposed difference in CB and VB symmetry between CdSe and PbSe QDs, should be reflected in noticeable differences in carrier-phonon coupling and carrier dynamics in these materials. In the next subsection we discuss how close these expectations are to experimental observations.

2.1.3 *Experimental Rates of Carrier Relaxation in Quantum Dots: Phonon Bottleneck Problem*

The phonon-mediated relaxation of electrons and holes in QDs is the focus of many experimental efforts [Nozik (2001); Ellingson *et al.* (2005); Murphy *et al.* (2006); Schaller and Klimov (2004); Schaller *et al.* (2005a, 2006); Klimov (2006); Ellingson *et al.* (2002); Petta *et al.* (2005); Crooker *et al.* (2002); Peterson and Krauss (2006)]. Throughout these studies, one of the most important and versatile tools available to the experimentalist has been ultrafast transient absorption (TA) pump-probe techniques [Klimov (2004)]. TA is a nonlinear optical method which allows for the monitoring (in both time and spectral domains) of absorption changes caused by photoexcitation. In the ultrafast TA experiment, nonequilibrium charge carriers are rapidly injected into a material with a short, typically subpicosecond pump pulse. Absorption changes associated with photogenerated carriers are monitored with a second short probe pulse which can be generated by a tunable, monochromatic, or broadband light source. By monitoring the dynamics of intraband relaxations, one can directly evaluate the rates of these transitions [Klimov (2000)].

TA experiments report fast subpicosecond carrier energy relaxation in bulk semiconductor crystals and attribute the dynamics to the Fröchlich interactions with longitudinal optical (LO) phonons [Klimov *et al.* (1995); Prabhu *et al.* (1995)]. However, the spacing between discreet electronic levels of a QD, clearly seen in optical spectra, is many times larger than the energy of an optical phonon. This spacing should dramatically slow down the relaxation and lead to a phonon bottleneck. The first prediction of slowed cooling at low light intensities in quantized structures was made by Boudreaux et al. [Boudreaux *et al.* (1980)]. They anticipated that cooling of carriers would require multiphonon processes when the quantized electronic levels were separated in energy by more than phonon energies. Because relaxation through several phonons is very inefficient due to the requirement of a resonance between the level separation and the total energy of multiple phonons, multiphonon relaxation time could be > 100 ps, a thousand times slower than those in bulk materials.

It is quite remarkable that there are so many reports that both support [Guyot-Sionnest *et al.* (1999); Mukai and Sugawara (1998); Murdin *et al.* (1999); Adler *et al.* (1998); Klimov *et al.* (1999); Ispasiou *et al.* (2001); Yu *et al.* (1996); Sugawara *et al.* (1997); Ellingson *et al.* (2002)] and contradict [Lowisch *et al.* (1999); Li *et al.* (1999); Gontijo *et al.* (1999);

Woggon *et al.* (1996); Sosnowski *et al.* (1998); Heitz *et al.* (1998)] the prediction of the existence of a phonon bottleneck in QDs. One element of confusion is that theory predicts an infinite relaxation lifetime of excited carriers for the extreme, limiting condition of a phonon bottleneck; thus, the carrier lifetime would be determined by nonradiative processes and band edge PL would be absent. Although some of these publications report relatively long hot-electron relaxation times (tens of picoseconds) compared with what is observed in bulk semiconductors, the results are reported as not being indicative of a phonon bottleneck because the relaxation times are not excessively long and band edge PL is observed. Since the phonon bottleneck in QDs could facilitate CM it could therefore be important for solar-cell applications. On the other hand, carrier relaxation could compete with electron transfer to the leads. Thus, both processes CM and electron transfer have to be faster than carrier relaxation, in order to increase efficiency of solar-cells. Because CM is very fast (subpicosecond time-scale), Nozik [Boudreaux *et al.* (1980)] suggested a comparison of the rate of relaxation/cooling with the rate of electron transfer. If the time of carrier relaxation/cooling in QDs is greater than electron transfer, i.e. > 10 ps (about an order of magnitude greater than the relaxation rate in the bulk semiconductor), then the relaxation would be considered slowed down by a "phonon bottleneck." Under such conditions, photoinduced hot electron transfer can be competitive with electron relaxation processes in QDs.

In III-V semiconductor QDs, such as GaAs, InAs and InP, slow charge-carrier cooling with relaxation times from 10 ps to 1 ns is observed [Yu *et al.* (1996); Sugawara *et al.* (1997); Murdin *et al.* (1999); Ellingson *et al.* (2002); Heitz *et al.* (1997); Adler *et al.* (1998)], and indeed, indicates a phonon bottleneck. The studies of CdSe and other colloidal QDs of the II-VI type mostly found two relaxation time scales whose relative weights depended upon the ligands used to passivate the QD surface [Yu *et al.* (2004); Guyot-Sionnest *et al.* (1999); Klimov *et al.* (2000a); Achermann *et al.* (2003); Ispasiou *et al.* (2001); Mohamed *et al.* (2001)]. Klimov et al. were the first to observe very fast 0.3 ps $1P_e$ to $1S_e$ intraband transitions in CdSe Qds [Klimov and McBranch (1998)]. These were attributed to an Auger process for electron relaxation that bypassed the phonon bottleneck. Other experiments [Guyot-Sionnest *et al.* (1999)] were conducted with CdSe QDs coated with different capping molecules (TOPO, thiocresol, and pyridine) exhibiting different hole-trapping kinetics. The rate of hole trapping increased in the order of TOPO, thiocresol, and pyridine. The results generally show a fast relaxation component (1-2 ps) and a slow relaxation component (~ 200 ps). The slow component dominates the data for the pyridine cap, which is attributed to its faster hole-trapping kinetics.

In another experiment by Klimov et al. [Xu *et al.* (2002); Klimov *et al.* (2000a); Klimov (2000)], they comparedcore-shell CdSe/ZnS Qds (expected to have no hole traps)and CdSe-pyridine QDs (expected to have strong hole traps). The observed relaxation times were 0.3 ps and 3 ps respectively, which seems to confirm the role that Auger recombination plays in the intraband relaxation. However, unlike the previous experiment, no slow component was observed. One possible explanation for the discrepancy is the smaller size of the nanocrystals in the second study, enhancing the role of the surface in the relaxation process.

Carrier relaxation in PbSe QDs from the 1P to the 1S electronic states also demonstrates very fast timescales of 2-6 ps [Harbold *et al.* (2005a,b); Schaller *et al.* (2005d,c)]. Even more surprisingly, the experimentally detected carrier relaxation rates in PbSe and CdSe QDs are faster in smaller QDs where the energy spacing is larger [Harbold *et al.* (2005a); Schaller *et al.* (2005c)]. Thus, there are many investigations in which phonon bottlenecks were apparently not observed and hot-electron relaxation was slowed only slightly compared to relaxation in bulk materials.

So far, three possible explanations of the absence of the bottleneck phenomenon have been suggested: carrier relaxation through surface states, through Auger recommendation [Schaller *et al.* (2005c)], and through multiphonon processes. However, fast relaxation can be facilitated by the manifold of surface states due to the high surface-volume ratio and imperfect surface passivation leading to impurities and defects in small QDs. In CdSe QDs it was shown that electron dynamics depend strongly on surface ligands while hole relaxation remained almost unaffected by changes in passivation [Guyot-Sionnest *et al.* (2005); Klimov *et al.* (2000a)]. Yet the role of surface states is not clear because the experimental control of QD shape and surface remains a challenge.

The Auger electron-hole scattering mechanism described by Efros et al. [Efros and Rosen (2000)] relies on a much higher density of hole states than electron levels, such as in CdSe QD, where the effective mass of a hole is ~ 3 times larger than the effective mass of an electron [Klimov (2004)]. Due to the pronounced asymmetry between the CB and the VB, holes relax significantly faster than electrons. In turn, electrons are able to lose energy through coupling with dense hole states [Ellingson *et al.* (2005)], providing overall ultra-fast relaxation of carriers. Experiments in which Auger-type electron-hole energy transfer is intentionally obstructed (electrons are decoupled from holes by spatial separation) [Klimov *et al.* (2000a)] show an increase in relaxation time by an order of magnitude and support the idea that Auger-like mechanisms open up new relaxation channels.

However, it is still unclear how intraband relaxation occurs in materials where Auger-like mechanisms are inefficient. In fact, semiconductor materials of IV-VI group, such as PbSe, are thought to have nearly symmetric CB and VB and, correspondingly, similar effective masses of the electrons and holes [Wise (2000)]. As a result, the Auger relaxation processes are inefficient when compared to a nanocrystal such as CdSe with asymmetric CB and VB. In the case of PbSe QDs, one possibility is relaxation through a multi-phonon channel. Although relaxation through several phonons is very inefficient due to the requirement of a resonance between the level separation and the total energy of multiple phonons, localization of wavefunctions and strong nonadiabatic electron-phonon coupling in strongly-confined QDs make a multiphonon process probable [Ridley (1982); Ledebo and Ridley (1982); Schaller *et al.* (2005d)]. The observed temperature-activated behavior of relaxation rates in PbSe QDs [Schaller *et al.* (2005d)] indirectly support the possibility of a multi-phonon relaxation channel.

Indeed, the explanation for an absence of phonon bottlenecks in QDs has strong fundamental importance. This question should also be addressed for its practical significance. In a QD gain medium, for example, it is desirable that excited carriers relax rapidly to their lowest states from which radiative recombination occurs. The phonon bottleneck would therefore be likely to hinder the operation of light-emitting devices. On the other hand, in QD solar materials that rely on CM [Schaller and Klimov (2004); Ellingson *et al.* (2005); Schaller *et al.* (2005a); Murphy *et al.* (2006)], the existence of a phonon bottleneck allows CM to compete effectively with relaxation processes. To get a better understanding of the properties of charge carriers in confined systems, further experimental, theoretical, and numerical investigations in dynamic properties of QDs should be performed.

2.2 Simulations of Nonadiabatic Dynamics: Trajectory Surface Hopping (TSH) in Density Functional Theory (DFT)

Existing theoretical approaches primarily focus on static characterization of QD structure and spectra with a few efforts devoted to direct real-time modeling of experimental data. In this Section we present the numerical method developed specifically for calculating carrier dynamics in relatively large systems. It is based on a TSH approach implemented by DFT-based molecular dynamics, which we also refer to as time-dependent Kohn-Sham or TDKS theory. The TDKS theory [Marques and Gross (2004); Baer and Neuhauser (2004)] is a well-known method of treating explicit electron [Baer and Neuhauser (2005)] and electron-nuclear dynamics in real time. In the electron-nuclear dynamics aspect, the TDKS is a many-body DFT version of the quantum-classical mean-field, or Ehrenfest, approach [Negele (1982)]. In this approach, the many-body wavefunction is reduced to a system of single-particle wavefunctions subject to an effective potential which is an average of the adiabatic states. To include nonadiabatic molecular dynamics, the fewest-switches TSH approach [Tully (1990); Hammes-Schiffer and Tully (1994); Parahdekar and Tully (2005)] has also been implemented in the model. The TSH method is one of the most popular fully atomistic nonadiabatic molecular dynamics (NAMD) schemes [Craig *et al.* (2005); Li *et al.* (2006); Coker (1993); Prezhdo and Rossky (1997b); Prezhdo and Brooksby (2001)] because it captures the essential physics – including detailed balance [Parahdekar and Tully (2005)] – while remaining computationally simple. It can be viewed as a quantum master equation for electron dynamics where the state-to-state transition rates depend on time through coupling to explicit phonon dynamics. In this section we represent the TSH approach for DFT with the Kohn-Sham (KS) representation of the electron density which greatly expands the range of TDKS applications. We outline the TDKS formalism in Subection 2.2.1 and present the TSH extension in Subsection 2.2.2.

2.2.1 *Time-Dependent Kohn-Sham Theory (TDKS)*

The electron density in DFT-based molecular dynamics is written in the KS representation [Marques and Gross (2004); Baer and Neuhauser (2004); Tretiak *et al.* (2005)] as

$$\rho(x,t) = \sum_{p=1}^{N_e} |\varphi_p(x,t)|^2 \tag{2.7}$$

where N_e is the number of electrons and $\varphi_p(x,t)$ are single-electron KS orbitals. The evolution of the $\varphi_p(x,t)$ is determined by application of the

TD variational principle to the KS energy

$$E\{\varphi_p\} = \sum_{p=1}^{N_e}\langle\varphi_p|K|\varphi_p\rangle + \sum_{p=1}^{N_e}\langle\varphi_p|V|\varphi_p\rangle \tag{2.8}$$

$$+\frac{e^2}{2}\int\int\frac{\rho(x',t)\rho(x,t)}{|x-x'|}d^3xd^3x' + E_{xc}\{\rho\}. \tag{2.9}$$

The right-hand side of Eq. 2.8 gives the kinetic energy of noninteracting electrons, the electron-nuclear attraction, the Coulomb repulsion and the exchange-correlation energy functional that accounts for residual many-body interactions. Application of the variational principle leads to a system of single-particle equations [Marques and Gross (2004); Baer and Neuhauser (2004); Tretiak *et al.* (2005)]

$$i\hbar\frac{\partial\varphi_p(x,t)}{\partial t} = H(\varphi(x,t))\varphi_p(x,t),\ p=1,\ldots,N_e \tag{2.10}$$

where the Hamiltonian H depends on the KS orbitals. In the generalized gradient approximation [Perdew (1991)] used for the present simulation, E_{xc} depends on both the density and density gradient, and the Hamiltonian is written as

$$H = -\frac{\hbar^2}{2m_e}\nabla^2 + V_N(x) + e^2\int\frac{\rho(x')}{|x-x'|}d^3x' + V_{xc}\{\rho,\nabla\rho\}. \tag{2.11}$$

The time-dependent $\varphi_p(x,t)$ can be expand in the adiabatic KS orbitals $\tilde{\varphi}_k\,(x;R)$

$$\varphi_p(x,t) = \sum_{k=N_1}^{N_2} c_{pk}(t)\big|\tilde{\varphi}_k\,(x;R)\big\rangle. \tag{2.12}$$

Here limits of sum, N_1 and N_2, vary for the VB and CB. In ideal case, $N_1 = 1$ and $N_2 = N_e$ for the VB (hole states); $N_1 = N_e + 1$ and $N_2 = \inf$ for the VB (electron states). The difference, $N_2 - N_1$, determine the energy window we consider and the accuracy of simulations. States with the energy larger (smaller) than the energy of the initial excitation of electrons (holes) contribute negligibly to the carrier relaxation. To reduce numerical expenses, such states are not included into consideration. Using nonadiabatic (NA) wavefunction (2.12) the TDKS Eq. (2.10) transforms to an equation of motion for the coefficients

$$i\hbar\frac{\partial}{\partial t}c_{pk}(t) = \sum_{m}^{N_e} c_{pm}(t)\Big(\epsilon_m\delta_{km} + \mathbf{d}_{km}\cdot\dot{\mathbf{R}}\Big). \tag{2.13}$$

The NA coupling

$$\mathbf{d}_{km} \cdot \dot{\mathbf{R}} = -i\hbar \langle \widetilde{\varphi}_k \left(x; R\right) \left| \nabla_{\mathbf{R}} \right| \widetilde{\varphi}_m \left(x; R\right) \rangle \cdot \dot{\mathbf{R}} = -i\hbar \langle \widetilde{\varphi}_k \left| \frac{\partial}{\partial t} \right| \widetilde{\varphi}_m \rangle \quad (2.14)$$

arises from the dependence of the adiabatic KS orbitals on the nuclear trajectory and is computed most easily from the right-handside of Eq. (2.14) [Hammes-Schiffer and Tully (1994)].

The time-dependence in TDKS for electron-nuclear dynamics is due to ionic motion, making $V_N(x) \equiv V_N\left(x; \mathbf{R}(t)\right)$ dependent on time through the nuclear trajectory $\mathbf{R}(t)$. The prescription for $\mathbf{R}(t)$ constitutes the quantum backreaction problem. In order to define the backreaction, TSH uses a stochastic element that creates both classical trajectory branching [Tully (1990)] and detailed balance [Parahdekar and Tully (2005)]. The former mimics the ability of quantum mechanical wavepackets to split and evolve in correlation with different electronic states. The latter is essential for relaxation and leads to thermodynamic equilibrium.

2.2.2 *Fewest Switches Surface Hopping in the Kohn-Sham Representation*

TSH chooses an electronic basis. Classical trajectories correlate with the states of this basis and hop between the states. Preferably, the TSH basis is formed by adiabatic states [Tully (1990); Hammes-Schiffer and Tully (1994); Parahdekar and Tully (2005)], i.e. eigenstates of the Hamiltonian (2.11). While adiabatic forces for ground and excited electronic states as well as the NA coupling between the ground and excited states can be calculated in TDDFT [Marques and Gross (2004); Baer and Neuhauser (2004); Tretiak *et al.* (2005)], the NA coupling between excited electronic states has not been commonly used yet. Ref. [Craig *et al.* (2005)] performs TSH in the zeroth-order adiabatic basis using Slater determinants formed by adiabatic KS orbitals. At present, we use a further approximation by going from many-particle to single-particle representation: TSH is performed in the basis of the single-particle adiabatic KS orbitals. The single-particle representation is well-suited to studies of QDs because the electronic structure is well-represented by the independent electron and hole picture (see discussion in Subsection 2.1.1). Quantum confinement effects in QDs ensure that the electron and hole kinetic energies dominate the electrostatic interaction. As a result, even basic EFM theory provides a good description of the QD electronic structure.

TSH prescribes a probability for hopping between electronic states. The probability is explicitly time-dependent and is correlated with the evolution of ions. In the fewest switches TSH [Tully (1990)], the probability of hopping between states k and m within time interval dt equals

$$dP_{km} = \frac{b_{km}}{a_{km}} dt, \tag{2.15}$$

where

$$b_{km} = -2\, Re\big(a^*_{km} \mathbf{d}_{km} \cdot \dot{\mathbf{R}}\big); \quad a_{km} = c_k\, c^*_m. \tag{2.16}$$

Here c_k and c_m are the coefficients evolving according to Eq. (2.13). The hopping probabilities explicitly depend on the NA coupling $\mathbf{d}_{km} \cdot \dot{\mathbf{R}}$ defined in Eq. (2.14). If the calculated dP_{km} is negative, the hopping probability is zero. This feature minimizes the number of hops: a hop from state k to state m can only occur when the electronic occupation of state k decreases and the occupation of state m increases. To conserve the total electron-nuclear energy after a hop, the nuclear velocities are rescaled [Tully (1990); Hammes-Schiffer and Tully (1994)] along the direction of the electronic component of the NA coupling $\mathbf{d}_{km}$. The hop is rejected if a NA transition to a higher energy electronic state is predicted by Eq. (2.15) and the kinetic energy available in the nuclear coordinates along the direction of the NA coupling is insufficient to accommodate the increase in the electronic energy. The velocity rescaling and hop rejection give detailed balance between upward and downward transitions [Parahdekar and Tully (2005)].

The current simplified implementation of TSH makes the assumption that the energy exchanged between the electronic and select nuclear degrees of freedom during the hop is rapidly redistributed between all nuclear modes. Under this assumption, the distribution of energy in the nuclear mode directed along the NA coupling $\mathbf{d}_{km}$ is Boltzmann at all times, and the velocity rescaling plus hop rejection can be replaced by multiplying the probability (2.15) for transitions upward in energy by the Boltzmann factor. Elimination of the velocity rescaling gives great computational savings by avoiding explicit quantum backreaction and allowing one to use a predetermined nuclear trajectory in order to evolve the electronic subsystem.

In summary, the simplified version of fewest switches TSH is performed in the single-particle representation with the hop rejection replaced by multiplication of the TSH probability upward in energy by the Boltzmann factor. The NA electronic evolution Eq. (2.13) is performed using the ground state nuclear trajectory. This treatment of the electron and hole relaxation creates a sophisticated version of the quantum master equation with explicit time-dependent transition probabilities that respond to nuclear evolution and give correct short [Prezhdo and Rossky (1998); Prezhdo (2000); Luis (2003)] and long-time dynamics [Tully (1990); Hammes-Schiffer and Tully (1994); Parahdekar and Tully (2005)].

2.2.3 *Simulation Details*

The PbSe and CdSe QDs used in the present study are shown in Fig. 2.3. The initial geometries of the clusters were generated based on the bulk lattice of CdSe (wurzite) and PbSe (zinc blende) respectively. Then the QD structures were fully optimized so that the total energy of the system reached its minimal value at zero temperature. All simulations were carried out in a cubic cell periodically replicated in three dimensions as stipulated by the plane-wave basis and used in the DFT approach. To prevent spurious interactions between periodic images of the QD, the cell was constructed to have at least an 8 $\AA$ vacuum layer that separates the QD replicas from each other.

To keep the simulation feasible, surface ligands were not included. In the case of PbSe, it is justified by the fact that surface effects are small in the PbSe QD relative to other types of semiconductors due to the higher order of its lattice symmetry [Wise (2000)]. For the CdSe QD, we checked the effect of a surface reconstruction on the electronic structure (specifically, on the near-gap levels) using the core-shell cluster $Cd_{33}Se_{33}/Zn_{78}S_{78}$ and $Cd_{33}Se_{33}$ passivated by 9 molecules of trimethylphosphine oxide $OP(CH_3)_3$ (TOPO) that are utilized as a model of ligands, commonly used for CdSe passivation. The electronic structures of pure CdSe, core-shell, and CdSe with ligands were compared to find the surface and trap states near the energy gap of the CdSe QD. The structures shown in Fig. 2.3 (b) were used only for electronic structure calculations. The large number of atoms in these systems restricts their use in DFT dynamics simulations.

The optimized configurations of $Cd_{33}Se_{33}$, $Pb_{16}Se_{16}$, and $Pb_{68}Se_{68}$ were brought up to 300 K by molecular dynamics with repeated velocity rescaling. A microcanonical trajectory was generated for each cluster using the Verlet algorithm with Hellmann-Feynman forces. The classical treatment of ion motion is justified at room temperature due to the fact that the frequencies of the available vibrational modes are on the order of or less than kT. Run in the ground electronic state, the time step was 2 fs for PbSe and 3 fs for CdSe, producing 4 and 3.6 ps trajectories respectively. The structures shown in Fig. 2.3 (a) were chosen randomly from these trajectories and indicate that the clusters, especially the largest one,

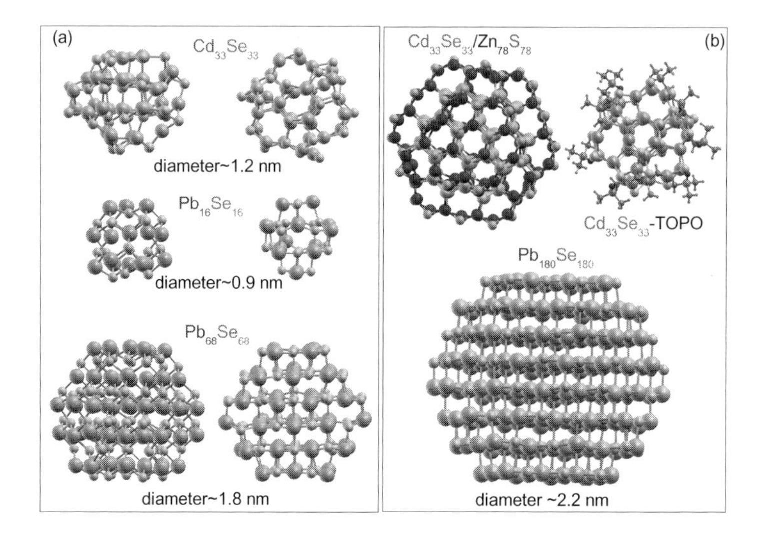

Fig. 2.3 Optimized structures of QDs used for electronic structure calculations and nonadiabatic dynamics (a) and used only for analyses of electronic structure and surface effects (b). The first and second panels in (a) show the (110)- and (111)-surface views respectively. Dark cyan, purple, and dark yellow correspond to Cd, Pb, and Se atoms in QDs; blue and pink-orange define Zn and S atoms in the shell of CdSe QDs; red, orange, grey and cyan represent O, P, C, and H atoms in the $OP(CH_3)_3$ ligands that model TOPO molecules passivating the CdSe QD. For color reference, turn to page 150.

preserve the bulk topology during the simulations. Fig 2.4 compares the relaxed atomic geometries of the smallest $Pb_{16}Se_{16}$ cluster at zero and at room temperatures with the initial ideal zinc blende bulk structure. Even at room temperature significant structural reconstructions of the QD due to relaxation were observed. Temperature-induced fluctuations distort the dot further but do not cause surface reconstruction or bond re-connectivity, thereby preserving the bulk bonding topology. The initial conditions for the NA dynamics were also sampled from a microcanonical trajectory for each QD. The transition dipole moments and oscillator strengths for excitations between KS orbitals were computed and used both to generate the optical absorption spectrum and to pick the most optically active excitations for the initial conditions of the NA runs. A 1 fs nuclear and a 10^{-3} fs electronic timestep were used for the NA dynamics calculations. The data shown in the figures below are converged by averaging over 500 runs.

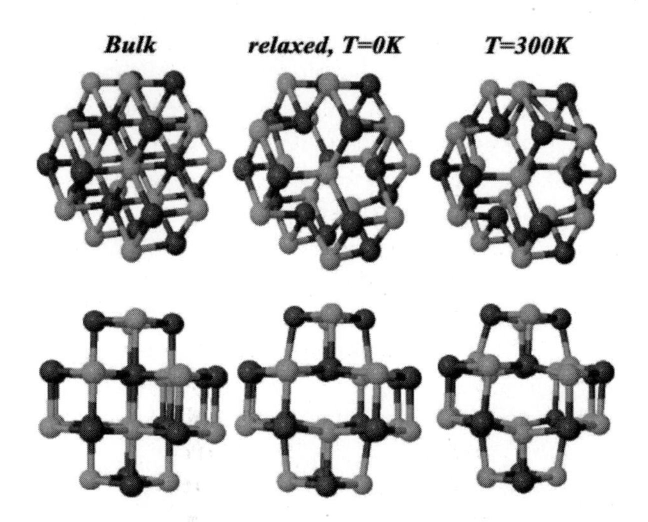

Fig. 2.4 Surface reconstruction and thermal effects on the geometry of the $Pb_{16}Se_{16}$ QD. The top row shows the (110)-surface view; the bottom shows the (111)-surface view. Pb atoms are black and Se atoms are yellow. The 32-atom QD preserves the bulk geometry in the very center of the QD even though significant relaxation at 0 K relative to bulk and thermal fluctuations in the structure at room temperature are observed at the surface. For color reference, turn to page 150.

Both the geometry optimization and adiabatic molecular dynamics were performed by plane wave DFT simulations implemented in the VASP [Kresse and Furthmüller (1996b); Kresse and Hafner (1994)] code, which is particularly efficient for semiconductor crystal band structure calculations. The core electrons were simulated using the ultrasoft Vanderbilt pseudopotentials [Vanderbilt (1990)], while all valence electrons were treated explicitly. The generalized gradient functional of Perdew and Wang (PW91) [Perdew (1991)] was used in order to account for the electron exchange and correlation effects. The simulations were performed using plane wave basis sets with over 50499 plane waves corresponding to an energy cutoff of 167.9 eV (12.34 Ry), which is large enough for the convergence of the total energy of these systems.

Typically, calculations of electronic structure based on pure DFT functional, such as PW91, lead to the electron self-interacting problem that results in much smaller simulated bandgaps than those in real systems. Implementation of some portion of the Hartree-Fock (HF) exchange in addition to the pure DFT exchange term usually helps to overcome this problem and improve calculated results. For this purpose, the DFT method implemented using the localized gaussian basis set and Gaussian-03 software package [Frisch *et al.*] was applied to the pure $Cd_{33}Se_{33}$ QD and to the capped ones, $Cd_{33}Se_{33}/Zn_{78}S_{78}$ and $Cd_{33}Se_{33}/OP(CH_3)_3$. The results obtained were compared to analogous calculations based on the VASP DFT code. In these simulations, we focused on how well various models of density functionals reproduce the experimental energy gap for each QD at zero temperature. We employed the most commonly used functionals, specifically the gradient-corrected functional Perdew, Burke, and Ernzerhof functional (PBE) and two hybrid functionals, the Becke 3-parameter hybrid functional (B3LYP) and the hybrid half-and-half functional (BHandH). B3LYP contains 20% of the HF exchange and BHandH 50%. The LANL2DZ basis set was used for all Gaussian calculations and is typically used for quantum-chemical simulations of inorganic systems and heavy atoms.

2.3 Results and Discussion

The reported state-of-the-art time-domain *ab initio* simulations of the charge relaxation processes in the PbSe and CdSe QDs provide novel and important details that were not accessible previously. In addition to electronic band structure which are available from the traditional simulations, our approach implicitly includes surface reconstruction effects. Also the NA approach directly probes relaxation mechanisms and timescales, identifies phonon modes that are responsible for the relaxation, and allows for direct comparison with the corresponding time-resolved experimental data.

2.3.1 *Electronic Structure at Zero Kelvin: Effects of Surface Reconstructions and Passivation*

As was discussed in Subsection 2.1.3, electronic structure and the symmetry of the density of states in particular determine the mechanisms of carrier relaxation in QDs. To address this issue we study the electronic structure of QDs, focusing on effects that surface reconstruction brings to the density of states of each QD.

Because the CdSe QD is affected more than the PbSe QD by processes taking place on their surfaces (see discussion in Subsection 2.1.2), we first examine the electronic structure of the optimized pure $Cd_{33}Se_{33}$ cluster and CdSe with two different cappings: a shell of ZnS semiconductor structure and nine molecules of trimethylphosphine oxide $OP(CH_3)_3$. Fig 2.5

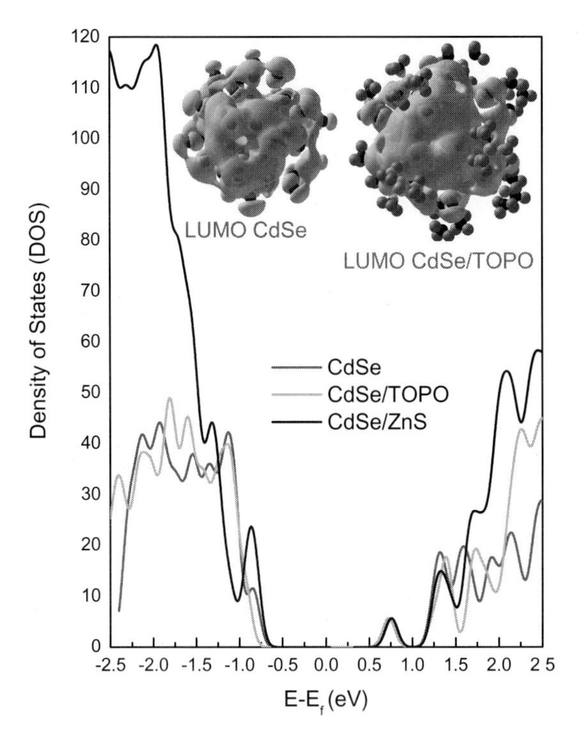

Fig. 2.5 Density of electron and hole states (DOS) of the pure $Cd_{33}Se_{33}$, $Cd_{33}Se_{33}$ with trimethylphosphine oxide (TOPO ligands) passivation, and core-shell $Cd_{33}Se_{33}/Zn_{78}S_{78}$ QDs at $T = 0$ K. Calculations are done by PW91 functional with plane-wave basis sets. The horizontal axis shows the energy of the states with respect to the Fermi level (E_f). The nearly accurate coincidence of energy gaps of all three structures proves the absence of trap states in the DOS of the pure CdSe. Thus, surface reconstruction of the bare CdSe QD leads to saturation of dangling bonds and, consequently, elimination of the trap states in the gap with the same efficiency as ligand passivation and ZnS coating. Inserts depict the partial charge densities of the lowest unoccupied molecular orbitals (LUMO) of the pure CdSe and CdSe/TOPO, revealing a similar character to both wavefunctions and their localization on the QD rather than the surface atoms.

compares the DOS for these three QD structures and shows very good agreement between the energy gap values of all these systems. This means that the bare CdSe QD does not have trapping states in the gap. Trapping states usually occur due to the dangling bonds on the surface of the QD. Using a semiconductor ZnS shell or molecules of trimethylphosphine oxide as a QD capping helps to saturate dangling bonds and remove trap-states from the gap. Our simulations demonstrate that the surface reconstruction due to the geometry optimization also saturates dangling bonds in the case of the bare CdSe QD. Previous DFT calculations of electronic structure of CdSe clusters [Puzder *et al.* (2004)] have also shown that geometry optimization of the surface leads to the compensation of dangling bonds and to the complete opening of the simulated gap in CdSe dots. Based on experimental results, it was proposed that surface reconstructions could only partially saturate dangling bonds. For example, it was reported that Se atoms on the surface come much closer to each other than they do in a bulk structure in order to impregnate the dangling bonds, thereby forming Se-Se bridges [Kasuya *et al.* (2004)].

Our analysis of the optimized geometries of pure CdSe and CdSe/TOPO QDs shows the presence of analogous Se-Se bridges in both structures. In addition, surface Cd atoms also reform their bonds, placing themselves closer to each other. For example, in both bare and TOPO-capped QDs the distance between the two closest Se atoms on the surface is ~ 3.9 Å compared to ~ 4.5 Å between the nearest Se in the center of the QDs. For the nearest Cd atoms, the distance is ~ 3.5 Å when they are on the surface and ~ 4.45 Å when they are in the center of the QD. Because Cd-Cd and Se-Se distances on the surface decreasesimilarly in both bare and passivated QDs, the surface in the pure CdSe QD experiences similar reconstructions as those in the passivated QD. This allows us to conclude that the electronic structure of the optimized bare CdSe QD includes most of the features of the passivated QDs, which are typically used in experiments. Thus, a bare optimized QD is a good model for simulations of the electronic properties of QDs.

Functional	CdSe	CdSe/TOPO
BHandHLYP	4.61	4.75
B3LYP	2.80	2.94
PBE	1.73	1.86
PW91	1.58	1.65
experiment	2.99	

The VASP-calculated zero temperature energy gaps are 1.58 eV for the pure $Cd_{33}Se_{33}$, 1.65 eV for the CdSe/TOPO QD, and 1.57 eV for the core-shell CdSe/ZnS QD. Each value is underestimated by roughly 1.2-1.3 eV. This is the standard systematic error for exchange-correlation functionals such as PW91 based on the adiabatic local density approximation (LDA) and gradient corrected functionals (GGA). The source of this failure is well-known; it is attributed to the lack of a derivative discontinuity in the LDA and GGA, stemming ultimately from the incomplete elimination of self-interaction by these functionals. Long-range nonlocal and nonadiabatic density functional corrections (such as hybrids that include a portion of the exact Hartree-Fock exchange) are routinely used to remedy the problem. As seen in Table 2.2 and Fig. 2.6, hybrid functionals lead to a consistent

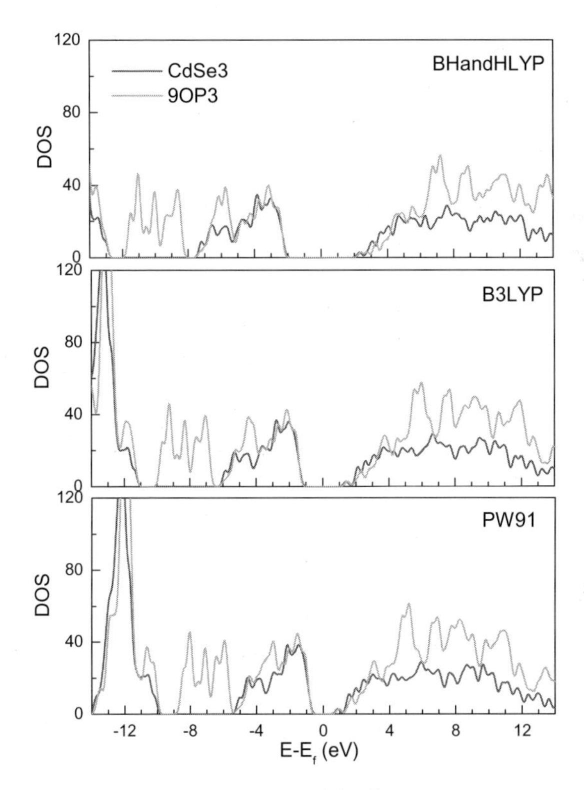

Fig. 2.6 Density of electron and hole states (DOS) of the pure $Cd_{33}Se_{33}$ and $Cd_{33}Se_{33}$ with trimethylphosphine oxide passivation at $T = 0$ K, calculated based on hybrid DFT functionals. The horizontal axis shows the energy of the states with respect to the Fermi level (E_f). The larger the amount of HF exchange present in the functional, the larger the energy gap in QDs. The overall structure of DOS does not depend on the functional. For excitation energies larger than double-bandgap energy ($2E_g$), the role of ligand states is important due to their dominated contribution to the DOS of the system. For example, ligand states of CdSe/TOPO completely occupy the VB sub-gap of the pure CdSe QD.

blueshift of the energy gap. Larger amounts of HF exchange present in the functional correspond to larger blueshifts. Overall, B3LYP provides the gap values closest to the experimentally observed ones in small $Cd_{33}Se_{33}$ and $Cd_{34}Se_{34}$ clusters of ~ 1.5 nm diameter [Kasuya *et al.* (2004)].

Fig. 2.6 compares the DOS of the bare CdSe and CdSe/TOPO that were calculated using different DFT functionals. It demonstrates that except for changes in the energy gap values, the overall character and structure of the DOS do not depend on the choice of a functional. Indeed, trimethylphosphine ligands as well as the ZnS shell contribute additional states to the DOS of the CdSe QD. That can be seen from the higher peaks at the energy roughly equal to and larger than double-gap energy ($2E_g$) of the CdSe/ZnS and CdSe/TOPO QDs, compared to those of the pure CdSe QD (Fig. 2.5 and Fig 2.6). For the excited energies larger than $2E_g$, capping states make an important contribution to the DOS of the system. Specifically, the first sub-gap in the VB of the CdSe QD is closing in the case of CdSE/TOPO. This can clearly be seen in Fig. 2.6. The states in this energy range originate from the trimethylphosphine molecules, and, of course, contribute to the charge relaxation at high-energy photoexcitations. Near the VB and CB edges, both types of capping result only in the slight shifting of levels with respect to the states of the bare CdSe QD. In this energy range, the VB and the CV of the CdSe dot are asymmetric, showing higher density of hole states (higher peaks in the DOS in Fig. 2.6) compared to the electron levels.

The DOS of PbSe QDs is represented in Fig. 2.7. This figure compares the electronic structure of three PbSe QDs of different size. The confinement effect is clearly seen: the smaller the diameter of the QD, the larger its gap. The energy gap of the smallest PbSe QD is $E_g = 1.24$ eV, $PbSe_{68}Se_{68}$ has $E_g = 1.15$ eV, and the energy gap of the largest $PbSe_{180}Se_{180}$ is $E_g = 1.03$ eV. These values, as is typical with DFT, are underestimated. Strong confinement also results in well-separated levels in the smallest cluster. However, for larger QDs that are still in the strong confinement regime, the discrete character of the electronic structure becomes less pronounced. The surface reconstructions that break the symmetry of the QD structure and lift the degeneracy of electronic levels rationalize the dense character of the DOS in strongly confined PbSe QDs.

Comparing the calculated DOS of both CdSe (Fig. 2.5) and PbSe (Fig. 2.7) QDs of 1-2 nm diameter, we can conclude that their electronic structure is not as discreet as earlier theoretical studies such as EMA [Efros and Efros (1982); Brus (1984)] have predicted. In these theories, electronic

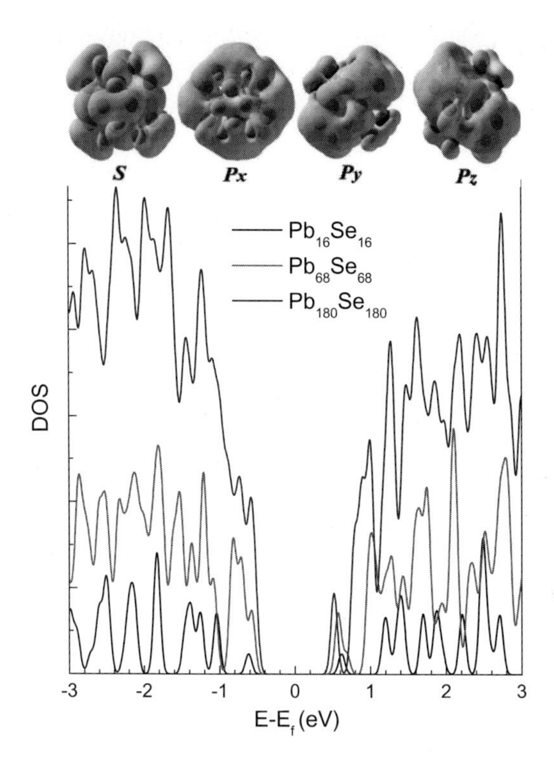

Fig. 2.7 Density of electron and hole states (DOS) of $Pb_{16}Se_{16}$, $Pb_{68}Se_{68}$, and $Pb_{180}Se_{180}$ QDs at 0 K, calculated with PW91 functional. The horizontal axis shows the energy of the states with respect to the Fermi level (E_f). The decrease in the QD diameter leads to an increase in the energy gap. The CB and VB show good symmetry for the smallest QD. For larger QDs, the hole states (VB) become slightly more dense than the electron states (CB). However, this asymmetry is much less pronounced than in the case of the CdSe QD (see Fig. 2.5). The insert depicts the charge densities of the lowest excited electronic states of $Pb_{16}Se_{16}$ whose space symmetry characterizes them as S and P_x, P_y, P_z states.

levels are highly degenerate (eightfold and sixfold degenerate ground states of PbSe and CdSe QD respectively). However, spin-orbit and Coulomb coupling, anisotropy, surface shape, and surface-ligand interactions should eliminate the degeneracy and reveal a more complicated multilevel character of the near gap states. Recent atomistic pseudopotential calculations [An *et al.* (2006)], together with our current simulations, demonstrate a more subtle electronic structure of both PbSe and CdSe QDs than what effective mass or tight-binding calculations have suggested previously. Such a multi-level sub-band structure, instead of very discreet, is also supported by photobleaching data. The observed photoluminescence blinking in single CdSe [Shimizu *et al.* (2001); Kuno *et al.* (2001)], CdTe [Shimizu *et al.* (2001)], and PbS [Peterson and Krauss (2006)] QDs demonstrates the highly nonexponential decay of probability density for a given on or off time. This indicates that the photogenerated exciton is not transitioning between a single bright state (on) and a single dark state (off) but is instead distributed to multiple on/off states [Efros and Rosen (1997); Peterson and Krauss (2006)].

Interestingly, the separation of the states lying at the very edges of CB and VB from the main manifold (S-peak in EMA notations) is more pronounced with electrons than holes for both CdSe and PbSe QDs. Comparing the DOS of the pure CdSe and capped CdSe QDs, we have shown that the lowest distinct peak in the CB is not the trap but the QD core state instead. The partial charge densities of this state for CdSe and CdSe/TOPO QDs, as presented in Fig. 2.5, also show that the wavefunction of this state is localized on the dot rather than on the surface atoms and ligands. By analogy and based on the relatively large numeric values of energy gaps in PbSe QDs, we assume that the lowest distinct peak in the DOS of PbSe QDs does not originate from the dangling bonds. Thus, there are no trap states in the gap of PbSe QD due to surface reconstructions. The electron densities of the four vacant orbitals closest to the edge of the CB of the smallest PbSe QD are represented in the top panel of Fig. 2.7. Surprisingly, despite the very small size of the QD, the wavefunction shapes of these states exhibit roughly S- and P-symmetries of envelope functions as predicted by EMA models [Efros and Efros (1982); Brus (1984)]. Indeed, they are significantly modified by the local atomic structure. The S- and P-symmetries of the analogous states of the larger QDs and the corresponding hole states at the edge of the VB are less pronounced.

Another important result is that the DOS of the smallest $Pb_{16}Se_{16}$ cluster demonstrates a strong symmetry between its VB and CB, showing nearly equally distributed electronic and hole states with respect to the Fermi Energy. This symmetry is slightly perturbed in larger QDs (Fig. 2.7). Although hole states in larger PbSe QDs show slightly higher density than electron ones, this asymmetry is not as dramatic as the case of CdSe (Fig. 2.5). This suggests a nearly equal effective mass of carriers in PbSe QDs and agrees well with recent DFT-based band structure calculations of bulk PbSe [Albanesi *et al.* (2005)].

2.3.2 *Evolution of Electronic Structure upon Temperature*

Now we consider the electronic properties of CdSe and PbSe QDs at room temperature. As a result of the thermal motion of ions, each individual molecular level of the QD experiences a quasi-chaotic change in energy. The evolution of the density of states for each QD, averaged over 500 different configurations randomly obtained at 300 K, is shown in Fig. 2.8.

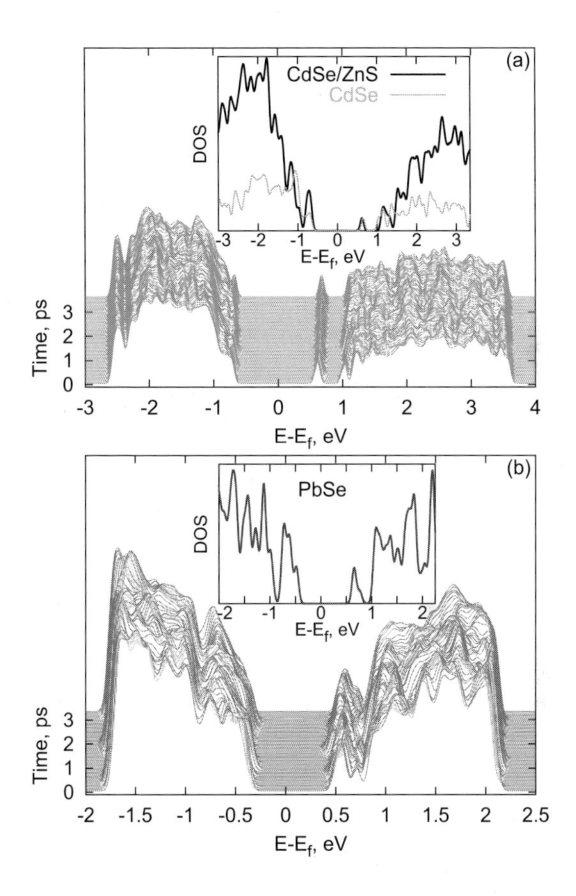

Fig. 2.8 Density of electron and hole states as a function of time along the MD run of (a) $Cd_{33}Se_{33}$ and (b) $Pb_{68}Se_{68}$ at room temperature. The CdSe states show a noticeable asymmetry across the gap, with the hole states having a higher density. This asymmetry is much less pronounced in PbSe. Inserts represent density of states (DOS) at $T = 0$ K. Dark blue line represents core-shell $Cd_{33}Se_{33}/Zn_{78}S_{78}$ QD, cyan line shows $Cd_{33}Se_{33}$ QD, and dark purple shows $Pb_{68}Se_{68}$ QD. The coincidence between CdSe DOS and DOS of the core-shell QD near their energy gaps demonstrates the absence of surface states in the gap of CdSe despite the uncapped surface of the CdSe dot. For color reference, turn to page 150.

The z-axis gives the state density as a function of energy and time. For both PbSe and CdSe QDs, the overall CB and VB energies do not acquire fluctuations larger than 0.1 eV in amplitude. The averaged energy gaps obtained at room temperature are $E_g = 1.3$ eV for $Cd_{33}Se_{33}$ and $Eg = 0.9$ eV for $Pb_{68}Se_{68}$, showing a decrease in gap with temperature for both QDs.

For comparison, the DOS of both QDs at $T = 0$ K is shown at the inserts of Fig. 2.8. The first peak at the CB, attributed to the so-called S_e state according to the EMA representation and originating from s-type electrons of Se atoms, is clearly seen for both QDs at zero and room temperatures. The next peaks in the CB can be attributed to the expected P_e, D_e, and so on, levels of electrons. For both QDs, the edge of the VB is significantly smoothed by the thermal motion of atoms, showing the vanishing of the small sub-band gaps in the VB, which are present at $T = 0$ K. Thus the VB has a more complicated character than the CB due to the mixing between its sub-bands. Nonetheless, there is an explicit correspondence between the main peaks at the VB at $T = 0$ K, as well as at $T = 300$ K for both types of QDs. For example, the first two main peaks in the VB of the PbSe QD can be assigned as S_h and P_h hole states. For CdSe the first three peaks in the VB can be associated with $1S_{3/2}$, $1P_{3/2}$, and $2S_{3/2}$ hole states in accordance with notations based on EMA [Ekimov *et al.* (1993)] (see discussion in Subsection 2.1.1). We use these notations later for assigning optical transitions in the simulated absorption spectra.

Interestingly, thermal smoothing of the VB is most pronounced for the smallest PbSe QD. Fig 2.9 illustrates the DOS of $Pb_{16}Se_{16}$ at zero and at room temperature for two random snapshots of MD trajectories at time $t = 2.24$ ps (green line) and $t = 3.50$ ps (blue line). At $T = 0$ K, the first peak S_h at the edge of the VB and the first peak S_e at the edge of CB are nearly equally separated by sub-band gaps from the manifold of other states in the bands. The next peaks, corresponding to P_h (VB) and P_e (CB) are also well-pronounced and exhibit symmetry with respect to each other. Because both P_e and P_h peaks have the threefold degeneracy, they are roughly tree times higher than the first S-peaks. At room temperature, thermally-activated oscillatory dynamics destroy the degeneracy and change the qualitative character of the VB. Thus, the S_h state experiences mixing with a portion of P_h states. This leads to the demolishing of the sub-band gaps that separate S_h and P_h peaks at zero K. For example, at time $t = 2.24$ ps (green line in Fig 2.9) S_h peak is still well-isolated, while at $t = 3.50$ ps (blue line in Fig 2.9) it becomes higher and broader. This

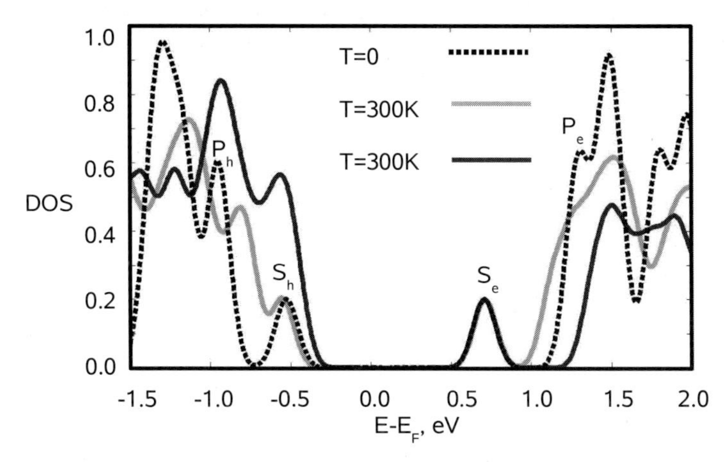

Fig. 2.9 Density of electron and hole states of $Pb_{16}Se_{16}$ at $T = 0K$ (dashes) and at $T = 300K$ (solid lines). DOS at ambient temperature is presented for two random snapshots of MD trajectory: at time $t = 2.24$ ps (green line) and at $t = 3.50$ ps (blue line). Thermally-activated oscillatory dynamics of electronic levels lead to breaking of the degeneracy, changing the qualitative character of the VB from gap-separated peaks to nearly continuous band. The energy and position with respect to the other band of the first peak in the CB stay almost independent of temperature and MD. However, in both the VB and CB the first two main peaks are recognized and can be attributed to S and P peaks. For color reference, turn to page 151.

change is caused by the closeness of two states formerly belonging to the P_h peak in energies to S_h. In contrast, the S_e peak (CB) stays well-separated from P_e, although P_e also spreads due to the decreasing of the degeneracy. Because HOMO and LUMO are localized on different atoms (Se and Pb), thermal motion of Se and Pb activate different phonon modes; the VB and CB have different electron-phonon coupling. In addition, because Se atoms lighter than Pb, the thermal motion of ions affects the edge of the VB much stronger than the CB in PbSe QDs, and this effect is more pronounced in small systems. This points to the stronger hole-phonon coupling that might result in faster relaxation in small dots.

Thermal motion reduces the discreetness of the electronic structure of both QDs, increasing splitting between nearly isoenergetic levels and further lifting the degeneracy. The overall shape of the thermally activated DOS follows the trend of the zero temperature DOS in all QDs we studied. Thus, in the CdSe QD the peaks near the edge of the VB represent a larger density (at least twice or more wider) than the edge peaks at the CB as shown in Fig. 2.8 (a). Consequently, the DOS of CdSe demonstrates a very

	holes	holes	electrons	electrons
QD	$1E_g$	$0.5E_g$	$0.5E_g$	$1E_g$
$Cd_{33}Se33$	34	42	179	100
$Pb_{16}Se16$	70	172	168	92
$Pb_{68}Se68$	28	21	38	27
$Pb_{180}Se180$	8	16	28	12

asymmetric shape due to the more dense VB compared to CB, associated with heavy holes and light electrons. The PbSe, in contrast, displays relatively symmetric character in its DOS. Although PbSe's holes show slightly higher density than electrons, this asymmetry is not as strong as in the case of CdSe. This supports the suggestion of nearly equal effective masses of carriers in PbSe QDs. Table 2.3 quantitatively advocates these conclusions, showing the splitting between states in the VB and CB averaged in the energy interval $0.5E_g$ and $1E_g$ starting from HOMO and LUMO for holes and electrons respectively. The relatively dense electronic spectra, their thermal fluctuations, and the difference in the symmetry between CB and VB of these two types of QDs are expected to have a significant impact on the relaxation processes in these systems.

2.3.3 *Optical Spectra of CdSe and PbSe Quantum Dots*

In strongly confined QDs such as those investigated here, the Coulomb interaction between a photoexcited electron and hole is much smaller than the quantum confinement energy (see discussion in Subsection 2.1.1). Thus, we can analyze the optical excitation approximately within a single-particle picture and treat its absorption spectra as the sum of independent contributions from the electron-hole transition dipole moments. Note that this is in contrast to SWCNTs where the electron-hole interaction is strong and dramatically affects energies of optical transitions, requiring many-body calculations to be performed (see Subsection 4.1.1).

Fig. 2.10 represents the optical absorption spectra of the CdSe, CdSe/ZnS, and PbSe QDs at room and zero temperatures. At 300 K, instead of reproducing the individual time slices, we represent an average over 500 single runs of MD trajectories. The comparison of results at room and zero temperatures demonstrate the quite similar main structure of spectra. In both CdSe and PbSe QDs, the thermal motion of ions broadens and

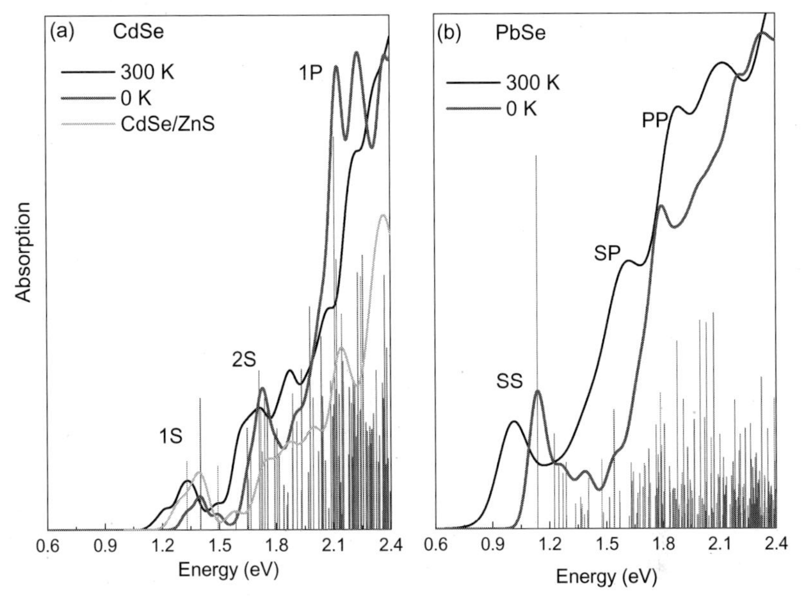

Fig. 2.10 Electron absorption spectrum averaged over the MD run at room temperature (black solid line) and at zero temperature (red lines) for (a) $Cd_{33}Se_{33}$ and (b) $Pb_{68}Se_{68}$. For comparison, the spectrum of core-shell $Cd_{33}Se_{33}/Zn_{78}S_{78}$ QD at $T = 0$ K is presented on (a) by the cyan line. The red vertical lines determine the electron-hole transitions with oscillator strength proportional to the line length. Both the pure and core-shell CdSe QDs demonstrate the first three peaks which can be attributed to 1S, 2S, and 1P transitions. In PbSe, the most pronounced peaks arise due to symmetric transitions across the gap, i.e. the SS and PP peaks. The spectrum also contains slightly less pronounced but nevertheless significant SP peaks due to asymmetric transitions. For color reference, turn to page 152.

shifts peaks to the red range of the spectra. The origin of this redshift and broadening is the quasichaotic time evolution of energies involved in optical transitions due to the thermally-activated vibrations or phonons. Despite the dense DOS, there are discreet transitions in the optical spectra of QDs that provide the most pronounced peaks. This is similar to the optical spectra of SWCNTs where only a few states are optically allowed among the dense manifold, while others are dark and do not contribute to the spectra (see Subsection 4.2.4).

The peaks in the QD absorption spectra can be attributed to transitions known from previous experimental and theoretical works (see Subsection 2.1.1). For both CdSe and CdSe/ZnS QD, the energies of the first three large peaks coincide with optical transitions between hole-electron states: $1S_{3/2}$-$1S_e$, $2S_{3/2}$-$1S_e$, and $1P_{3/2}$-$1P_e$, which we abbreviate as 1S, 2S, and

1P on the plot. The qualitative structure of the spectra is in agreement with the experimental spectra of small $Cd_{33}Se_{33}$ clusters [Kasuya *et al.* (2004)] and larger CdSe nanocrystals with a diameter of 2 nm [Klimov and McBranch (1998)].

In the PbSe QD, due to selection rules based on its DOS symmetry, the strongest peaks in the spectrum correspond to the symmetric transitions across the energy gap, i.e. $S_h - S_e$ and $P_h - P_e$, abbreviated as SS and PP peaks on the plot. The strong oscillator strength of the very first SS transition, which mostly corresponds to the HOMO-LUMO transition, supports the fact that the distinct state from the edge of the CB (see Fig. 2.8) is not the surface state of the QD. Remarkably, asymmetric transitions such as S-P, which are parity-forbidden by the EMA model but experimentally observed in 4-6 nm PbSe QDs [Harbold *et al.* (2005a)], are clearly present in the spectra of $Pb_{68}Se_{68}$ at both room and zero temperatures. At T=0 K these transitions are much weaker than SS and PP transitions. The analogous SP peak is also reproduced in all simulated spectra of PbSe QDs of different size.

Fig. 2.11 shows the absorption spectra of $Pb_{16}Se_{16}$, $Pb_{68}Se_{68}$, and $Pb_{180}Se_{180}$ QDs at zero temperature. In this figure, the energy of each electron transition is scaled with respect to the energy gap of each QD and represented as a ratio of the transition energy to the QD energy gap. The first peak in the spectra, which corresponds to the HOMO-LUMO transition (SS), is placed at 1 for all QDs. Thus, Fig. 2.11 demonstrates the ratio between energies of SS, SP, and PP peaks in the spectra for different PbSe QDs. The analogous representation of experimental spectra of PbSe, PbS, and PbTe QDs has been reported in Ref. [Wehrenberg *et al.* (2002); Murphy *et al.* (2006)]. Their data show that the normalized confinement energy for the second PP peak remains constant for a given material, independent of the QD size and temperature; we also find this value to be constant against QD size in PbSe. Fig 2.11 also demonstrates that symmetric transitions across the gap, i.e. the SS and PP peaks, and asymmetric transitions SP peaks are roughly the same for all three QDs. This indicates that although the transition energies are size-dependent, the overall structure of the spectra does not depend on the QD diameter, allowing for clear identification of SS transitions, SP transitions, and so on. Remarkably, these ratios coincide with experimental data [Harbold *et al.* (2005a)] that give $\sim$ 1.6 eV PP/SS energies and $\sim$ 1.3 eV SP/SS energies.

Since the energy splitting between the S- and P-states is slightly different for electrons and holes and is larger for the electrons than for holes

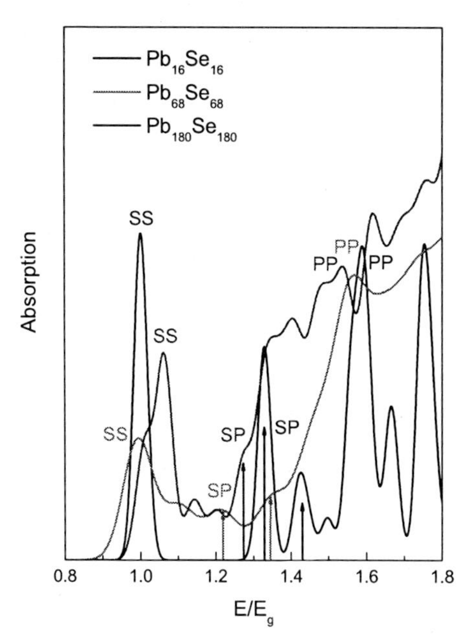

Fig. 2.11 Electron absorption spectrum of $Pb_{16}Se_{16}$, $Pb_{68}Se_{68}$, and $Pb_{180}Se_{180}$ QDs at T=0 K. The x-axis presents transition energy divided by the energy gap (E_g) of each QD. The first peak, corresponding to the HOMO-LUMO transition (SS), is placed at 1 for all QDs. The ratio between energies of symmetric transitions across the gap, i.e. the SS and PP peaks, as well as for asymmetric SP peaks, is roughly the same for all three QDs. The arrows indicate $S_h - P_e$ and $P_h - S_e$ transitions in each QD. These transitions belong to the SP peaks and have slightly different energies due to the more dense character of the VB than the CB in PbSe QDs.

(see Fig. 2.11), the S-P band contains two sub-peaks which are clearly pronounced in the spectra of $Pb_{16}Se_{16}$ and $Pb_{68}Se_{68}$ dots. The lower energy peak arises due to transitions between the P-states of holes and S-states of electrons. The higher energy sub-peak creates excitations from S-states of holes to P-electron states. For the larger $Pb_{180}Se_{180}$, the identification of the SP peak is complicated by the more dense and mixed character of its VB (see Fig 2.7).

The origin of this SP peak is still under debate. On the basis of the coincidence between the measured [Talapin and Murray (2005); Harbold *et al.* (2005a); Ellingson *et al.* (2005); Liljerothos *et al.* (2005); Schaller and Klimov (2004); Schaller *et al.* (2005d)] and $\mathbf{k} \cdot \mathbf{p}$ calculated [Allan and Delerue (2004); Ellingson *et al.* (2005); Kang and Wise (1997); Andreev and Lipovskii (1999)] transition energies, it was suggested that the three observed peaks in the PbSe absorption spectra originate from $S_h - S_e$, $S_h - P_e$ (or $P_h - S_e$) and $P_h - P_e$ transitions respectively. Thus, in this interpretation the second observed absorption peak corresponds to formally forbidden optical transitions PS and SP that violate the parity conservation rule. The $\mathbf{k} \cdot \mathbf{p}$ theory was modified to provide new selection rules for optical transitions and allow SP transitions. Specifically, these modifications include (1) a softening of the parity selection rule due to the anisotropy of the QD surface [Kang and Wise (1997)] and (2) inclusion of band anisotropy in the electronic structure calculations [Andreev and Lipovskii (1999)].

An alternative approach was suggested by Zunger [An *et al.* (2006)]. Based on the atomic-pseudopotential method, he concludes that the so-called SP peak is actually an optically allowed PP transition, while the third peak in the spectra, commonly attributed to PP, should be called a DD transition. This new result originates from the asymmetry between the VB and CB in this method, instead of symmetric bands obtained by EMA and $\mathbf{k} \cdot \mathbf{p}$ approaches (see discussion in Subsection 2.1.2). These results have recently been supported by scanning tunneling microscopy (STM) measurements [Liljerothos *et al.* (2005)]. However, such an interpretation of the mysterious SP peak does not explain the smaller intensity of this peak compared to the other two. It also does not elucidate the enhancement of the SP peak versus the suppression of the SS and PP peaks in the two-photon absorption spectra.

Our simulations provide roughly symmetric CB and VB in PbSe QDs, especially for the smallest $Pb_{16}Se_{16}$ QD, with only slightly heavier holes than electrons. This allows us to interpret the SP peak as an asymmetric transition between S-hole and P-electron states (or P-hole and S-electron). In our approach, the optical activity of SP transitions is rationalized by a breaking of both bulk structure symmetry and spherical-shape symmetry of the QD due to surface reconstructions. It is important to note that understanding the origin of these SP transitions is important to the understanding of carrier dynamics. Thus, Ellingson et al. [Ellingson *et al.* (2005)] have recently argued that because asymmetric transitions such as PS are expected to lead to strong exciton-phonon coupling, they would allow rapid relaxation of excited carriers via electron-phonon interaction.

2.3.4 *Active Phonon Modes*

Thermal fluctuations of electronic levels depend directly on the specific phonon modes coupled to these levels. Therefore, analysis of the thermal fluctuations of each individual molecular orbital for each MD run can identify phonon modes that are responsible for the relaxation of photoexcited carriers. Fig. 2.12 shows the spectral density of the phonon bath obtained by Fourier transform of the energies of the two electron-hole pairs for CdSe and PbSe QDs. These pairs are expected to be the most optically active

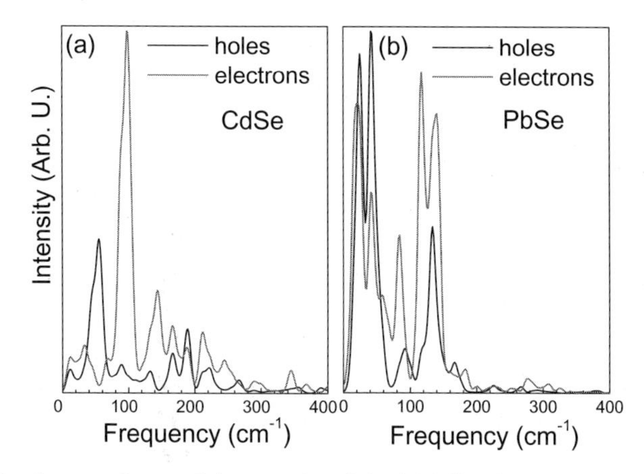

Fig. 2.12 Fourier transforms of the energies of the initially photoexcited states of holes (blue line) and electrons (magenta line) of $Cd_{33}Se_{33}$ (a) and $Pb_{68}Se_{68}$ (b). The peaks correspond to the thermally activated phonon modes that induce the relaxation of carriers to the edges of their bands. For both QDs, both charge carriers interact with low frequency phonons. Electrons couple to slightly faster phonons. For color reference, turn to page 152.

states with large transition dipole moments. The pairs should have energies roughly three times larger than the QD energy gap ($3 \times E_g$) and be nearly equally separated from the center of the bandgap. These conditions are chosen for the initial photoexcitation of electrons and holes in both types of QDs to correspond with experiments [Schaller and Klimov (2004); Klimov and McBranch (1998)].

The NA coupling is directly related to the second derivative of the energy along the nuclear trajectory. Therefore, those vibrational modes that most strongly modulate the energy levels create the largest coupling. Fig. 2.12 indicates that phonons with frequencies around 50 and 100 cm^{-1} dominate both electron and hole dynamics for PbSe and CdSe QDs. The modes with frequencies around 200-250 cm^{-1} are also clearly present in the spectra but with a smaller amplitude. Raman experiments [Krauss and Wise (1997a,b); Yoffe (2001)] report frequencies of $\sim$ 45 and $\sim$ 70 cm^{-1} associated with spheroidal acoustic modes and $\sim$ 215 cm^{-1} attributed to the lowest-order LO mode in PbS. WPbS QDs are supposed to have similar phonon modes to PbSe QDs. In CdSe QDs, the LO mode is detected at $\sim 210\ cm^{-1}$, and frequencies of $5 - 50\ cm^{-1}$ are attributed to acoustic vibrations [Mittleman *et al.* (1994); Yoffe (2001)]. In both types of QDs, the LO modes are close in energy to the bulk LO phonons and only slightly vary with QD shape and size [Mittleman *et al.* (1994); Krauss and Wise (1997a)]. Well-pronounced peaks in Fig 2.12 agree well with these experimental data. This allows us to attribute the calculated low-frequency (50 and 100 cm^{-1}) modes to the acoustic phonons and attribute modes with frequencies of 200-250 cm^{-1} to the LO phonons in each QD we study.

The LO mode in QDs is analogous to the G mode in SWCNTs. The G mode corresponds to the stretching vibrations along the carbon-carbon bonds [Fantini *et al.* (2004a)]. Similar to the G mode in bulk graphite or graphene, the G mode in tubes typically has a frequency of $\sim 1580\ cm^{-1}$ and is not sensitive to tube diameter and diameter. In this respect, the low-frequency acoustic mode in QDs can be associated with the RBM mode in SWCNTs.The RBM collective motion mode corresponds to the atomic vibration of the C atoms in the radial direction, as if the tube was breathing. Analogous to the acoustic mode in QDs, the RBM is very sensitive to the size of SWCNTs and is used to characterize nanotube diameters [Saito *et al.* (1998a)]. However, the exciton coupling with G vibrations is stronger than with RBM modes in SWCNTs [Gambetta *et al.* (2006a); Shreve *et al.* (2007a)] (see discussion in Subsection 4.3.5). In contrast, acoustic phonons in QDs are more pronounced in experimental Raman spectra than the LO modes [Mittleman *et al.* (1994); Krauss and Wise (1997a)]; as a result, they are expected to have stronger coupling with excited electronic states than LO phonons have. The difference in amplitudes between the LO and acoustic modes in Fig 2.12 also indicates a stronger electron-phonon coupling for acoustic phonons compared to LO phonons in both QDs. This can be rationalized by the fact that QD energy levels are more sensitive to acoustic modes than to optical modes with local atomic displacements. Acoustic modes are modulated by QD size and shape while atomic displacements in optical modes tend to average out due to thermal fluctuations. Such a conclusion is supported by earlier [Krauss and Wise (1997a); Mittleman *et al.* (1994)] and recent Raman, PL, and photon-echo experiments [Fernee *et al.* (2007); Morello *et al.* (2007)] of PbS, PbSe, CdSe, and CdTe QDs. From these experimental observations, it was generalized that significantly enhanced acoustic phonon coupling in nanocrystals is indicated by both emission peak shift and large temperature-dependent homogeneous line shape. Strong interaction with acoustic phonons is also responsible for ultrafast dephasing ($\sim$ 5 fs) in PbSe QDs. This dephasing was predicted [Kamisaka *et al.* (2006)] and determined experimentally from the homogeneous line width [Fernee *et al.* (2007)]. However, because our numeric approach allows us to consider the system only in its ground electronic state during *ab initio* MD, we cannot go beyond qualitative estimations of electron-phonon coupling in comparing our calculations with experimental observables, such as Huang-Rhys factor and Stokes shifts. In contrast, the ESMD approach used in SWCNTs provides these observables, and we compare them with available experimental data in Subsection 4.3.5.

In $Pb_{68}Se_{68}$, the amplitude and frequency of active phonon modes are similar for electrons and holes. This suggests that the relaxation dynamics of the two charge carriersshould be similar. Phonon modes in the smaller $Pb_{16}Se_{16}$ demonstrate the analogous symmetric coupling with electron and hole states [Kilina *et al.* (2007)]. In contrast, low frequency phonons coupled with electron states in CdSe differ noticeably from those coupled with holes. Thus, the CdSe electron states have a strong coupling with the 100 cm^{-1} mode, while holes are better coupled with 50 cm^{-1} phonons. This difference indicates the asymmetry between CB and VB in CdSe QDs. The preferable participation of the faster phonon modes ($\sim$ 100 cm^{-1}) should tend to speed the electron relaxation. However, this effect is counterbalanced by the lower density of electron states (see Fig. 2.8) compared to hole states in CdSe QD.The analysis of the relaxation dynamics presented in Subsections 2.3.5 and 2.3.6 establishes which of the two factors contributes the most.

2.3.5 *Phonon-Induced Electron and Hole Relaxation*

The relaxation dynamics in PbSe and CdSe QDs are detailed in Figures 2.13 and 2.14 and Table 2.4. According to the TSH method described above, the relaxation appears as a sequence of transitions between nearest or next nearest levels. The stronger the NA coupling between the states, the higher the probability of a hop and, consequently, the faster the overall relaxation. Table 2.4 compares the NA couplings averaged over all electron (hole) states involved in relaxation and averaged only over the neighboring states for different excitation energies in CdSe and PbSe QDs. Because the coupling between nearest states is nearly 1.7 times. larger than the total coupling for both electrons and holes, independent of their initial excitation energies, the main contribution to carrier relaxation should come from neighboring transitions. For both QDs, holes have stronger NA coupling than electrons, rationalizing a faster relaxation time for holes. However, this difference is more pronounced in CdSe than PbSe. For $3 \times E_g$ excitation energy, the NA hole coupling is 1.8 times larger than the electron coupling for CdSe, compared to a ratio of 1.3 for PbSe.

			$Cd_{33}Se_{33}$			$Pb_{68}Se_{68}$		
		$2E_g$	$2.5E_g$	$3E_g$	$2E_g$	$2.5E_g$	$3E_g$	$4E_g$
electrons	$d_{i,i+1}$ (meV)	14.6	14.8	16.7	20.4	23.6	23.8	27.8
	d_{total} (meV)	8.6	8.9	9.9	12.0	13.8	15.1	16.2
	rate K (ps^{-1})	0.8	1.0	1.1	0.7	1.1	1.3	1.4
	crossover τ (ps)	0.3	0.3	0.3	0.3	0.4	0.5	0.5
holes	$d_{i,i+1}$ (meV)	22.4	22.5	22.4	30.1	31.4	32.2	33.3
	d_{total} (meV)	13.2	13.3	13.2	17.7	18.4	19.0	19.5
	rate K (ps^{-1})	1.6	1.9	1.9	1.2	1.4	1.6	1.6
	crossover τ (ps)	0.4	0.5	0.5	0.25	0.3	0.3	0.2

Interestingly, higher excitations have a larger averaged coupling and a faster relaxation which is in perfect qualitative agreement with recent experiments [Harbold *et al.* (2005a)]. This is explained by a significant increase in the density of states for orbitals lying far from the band gap (see Table 2.3). For example, more states can contribute to the averaged NA coupling for an excitation with energy $4 \times E_g$ than with $2 \times E_g$. The exception is the hole relaxation in the CdSe QD where the increase in excitation energy only slightly changes NA coupling and relaxation rates. Because the VB is much denser than the CB in CdSe, hole states are dense even near the gap ($\sim 2 \times E_g$) which rationalizes the exception. Thus, including the additional high-energy levels ($\sim 3 \times E_g$) only slightly changes the contribution of these states to the averaged coupling and overall relaxation.

The results of phonon-mediated electron and hole dynamics for initial excitation with energy $\sim 3 \times E_g$ for both QDs are represented graphically in Fig. 2.13. This shows a three-dimensional plot of the product of DOS with the state occupations as a function of energy and time. As evidenced by the data, the carriers visit multiple states during the relaxation, but none of the intermediate states play any special role. By comparing the DOS in Fig. 2.8 with the population dynamics shown in Fig. 2.13, one observes for both electrons and holes of both QDs that the initial photoexcitation peak vanishes, only to reappear at the final states. Due to the smaller

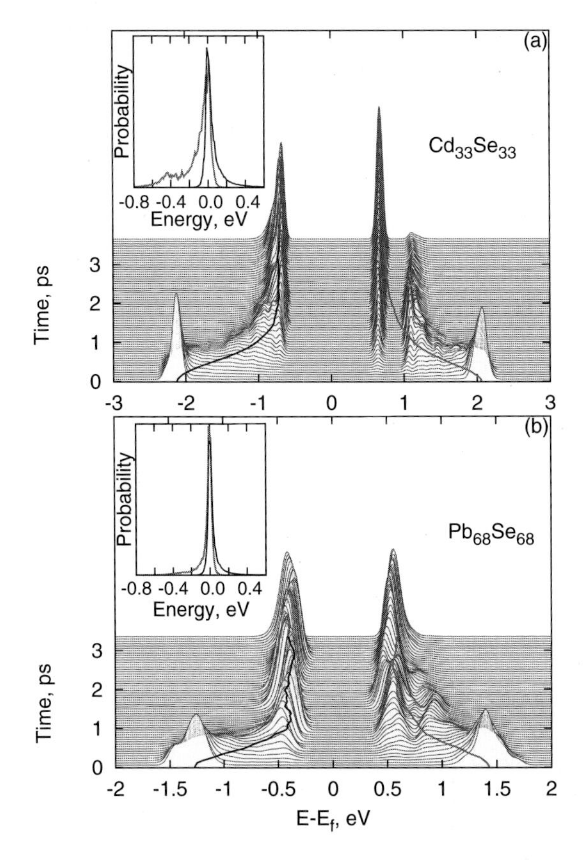

Fig. 2.13 Relaxation dynamics of electron and hole energies and populations for $Cd_{33}Se_{33}$ (a) and $Pb_{68}Se_{68}$ (b) QDs. Initially excited electron and hole time-dependent populations are computed by multiplying the DOS by the corresponding state occupations and averaging over 500 initial conditions. Blue and magenta lines correspond to average energies of electrons and holes respectively. The major part of the relaxation occurs within the first picosecond. In PbSe the holes evolve only slightly faster than the electrons, while for CdSe this difference is more noticeable. Insert: Probability density of the electron (blue line) and hole (magenta line) transitions as a function of the transition energy. While most hops occur down in energy and are responsible for the charge-phonon relaxation, some hops transfer energy from phonons to the charged particles. The energy exchanged during some transitions is greater than the LO phonon energy of 200 cm^{-1} (0.025 eV), providing a possible multi-phonon relaxation of carriers. For color reference, turn to page 153.

size of CdSe QDs, the very spare electron levels near the CB edgecause two distinct peaks to be present at the final time (3.5 ps) The peaks show $\sim$ 95% of the population on the LUMO and $\sim$ 5% on the LUMO+1. In PbSe QDs, the final peak in the CB (previously assigned as the $1S_e$) is broadened over several states from the very edge of the CB. The final hole population for both QDs shows a broad peak centered very close to the HOMO. Thus,carrier relaxation is nearly completed by 3.5 ps for both QDs.

Inserts in Fig. 2.13 illustrate the probability-density distributions for the electron and hole transitions as a function of the transition energy. For both QDs, these probabilities are normalized so that the area under the curve is equal to unity. The transitions occur both up and down in energy; however, transitions downward dominate as required by detailed balance. The overall energy flow from charged particles to phonons is responsible for the relaxation of the initial photoexcitation. The most likely transitions involve small amounts of energy that exceed the energy of a single phonon. For example, Fig. 2.13 shows the energy of a single phonon to be around 100 or 200 cm^{-1} (0.012 or 0.025 eV). However, up to 0.3-0.6 eV of electron energy can occasionally be lost in a single event. This mostly happens for electron transitions and is the most pronounced in the case of small PbSe (not shown) and CdSe QDs. The presence of transitions with energy at least 10 times larger than the energy of the most active phonons proves the concept of a multi-phonon relaxation channel in QDs. Thus, we conclude that the electron-phonon relaxation in QDs, particularly in a small CdSe dot, occurs primarily via the multiphonon mechanism [Schaller *et al.* (2005d); Harbold *et al.* (2005b)].

The average energy of both electrons and holes (magenta and blue lines in Fig. 2.13) plateaus after 1.5-2 ps, which indicates completeness of relaxation by that time. Importantly, average relaxation energy shows very distinct behavior at initial and final intervals of time. At the very beginning of the evolution, all charge carriers experience a short period when the transitions with energy loss and gain are equal. This short interval is associated with fine quantum effects and corresponds to development of correlations between a charge-carrier and a phonon subsystem. This regime corresponds to the Gaussian-like shape of energy relaxation dependence. Later, the energy relaxation has exponential decay with varying rates for electrons and holes. Overall, the relaxation of the electron and hole energies for both QDs shown in Fig. 2.13 has non-exponential character. This calculation agrees with the strongly non-Lorentzian line shapes observed experimentally [Califano *et al.* (2004)].

2.3.6 *Simulated Rates and Regimes of Carrier Relaxation*

To get a quantitative estimation of relaxation rates in CdSe and PbSe as well as compare Gaussian and exponential regimes of carrier relaxation, the derivative of the relaxation energy with respect to time is found in accordance with the formula:

$$\kappa(t) = \frac{d}{dt}\langle E_{carrier}\rangle(t) / \left(\langle E_{carrier}\rangle(t) - \langle E_{final}\rangle\right). \quad (2.17)$$

Here $\langle E_{carrier}\rangle(t)$ is the averaged electron (or hole) relaxation energy at specific time t. E_{final} stands for the average energy of a carrier at the end of the considered interval of time. Typically, $\langle E_{final}\rangle$ is very close to E_{LUMO} or E_{HOMO} for electrons and holes, respectively. Since energies of the bandgaps and electronic transitions are very different for different QDs, this approach allows the uniform comparison of relaxation rates κ for QDs with various electronic structures, independent of the absolute value of the energies.

Fig. 2.14 illustrates the dependence of relaxation rate κ on time for both QDs initially photoexcited to the energy triple of the bandgap of the QD. The overall shape of $\kappa(t)$ simulated in Fig. 2.14 assumes that the carrier relaxation rate grows nearly linearly at times smaller than charge carrier-phonon correlation time $t = \tau$ (crossover time). After the crossover time, the averaged rate does not change but fluctuates around some constant. Thus, a simple fitting of the rates has the following form

$$\kappa(t) = \begin{cases} -Kt, \, t < \tau; \\ -K, \;\; t > \tau, \end{cases} \quad (2.18)$$

The time interval before the crossover time corresponds to the Gaussian behavior of relaxation with rate constant K. After the crossover time, the relaxation has an exponential character with the same rate constant. The identity of rate constants in Gaussian and exponentional regimes deals with a significant NA electron-phonon coupling (several meV) in both QDs. Using the formula above, the fitting parameters for numerical rates are calculated for different initial excitation energies and summarized in Table 2.4.

The TSH approach describes the dynamics of carrier relaxation in QDs through the solution of Eq. 2.15 for Kohn-Sham orbital populations. Such a system of ordinary differential equations has a solution in the form of a linear combination of exponentials, only if coefficients are constant and real numbers, i.e. in Markov limit [Kondov *et al.* (2003)]. The nonexponential profile (the Gaussian form) of the population decay present in the current simulation is due to the explicit time-dependence of the relaxation rates in

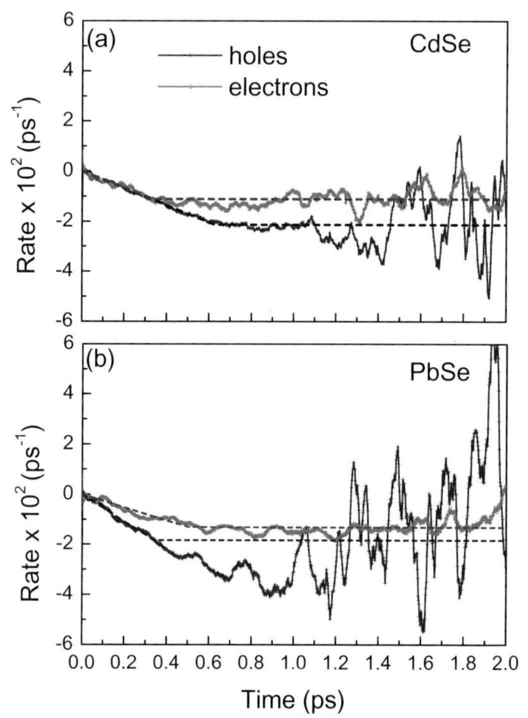

Fig. 2.14 The rates of relaxation of electrons (magenta) and holes (blue) for initial excitation energy $3 \times E_{gap}$ for (a) $Cd_{33}Se_{33}$ and (b) $Pb_{68}Se_{68}$ QDs. The rates were simulated according to the formula $rate = (dE/dt)/(E - E_{final})$, where E stands for energy of a carrier, E_{final} is the energy of the final state, and t is time. The dashed line corresponds to the fitting function $\kappa(t)$ satisfying Eq. 2.18. The time interval before the crossover time τ corresponds to the Gaussian behavior of relaxation with rate constant K. After the crossover time the relaxation has an exponential character with the same rate constant. The identity of rate constants in Gaussian and exponentional regimes deals with the significant electron-phonon coupling (NA coupling). In both QDs holes are faster than electrons. This observation is rationalized by a denser VB compared to CB and stronger NA coupling between hole states than between electron states. However, this difference is more pronounced for CdSe which has a more asymmetric band structure. For color reference, turn to page 153.

Eq. 2.15, providing a slow damping of the autocorrelation function as shown in Ref. [Kamisaka *et al.* (2006)]. The TSH correctly represents the vanishing time-derivative of the quantum-mechanical transition probability at zero time, which is manifested in the quantum-Zeno effect [Luis (2003)]. This determines the Gaussian relaxation component at earler times – smaller than the crossover time τ – before a system enters the exponential (Markov) regime of relaxation.

The Gaussian regime is associated with the phonon coherency in QDs. Initially, the electronic system is prepared in a pure state in which the electron-phonon correlation is maximal. Because of statistical approach, the electronic system is coupled to an ensemble of ionic subsystem replicas with a distribution of positions and momenta over a phase space. During a relatively short time interval, all replicas included in the ensemble move coherently in the same direction, preserving the phase relationship. The center of the replica distribution follows this motion. Later, for times larger than the crossover time τ, the difference in frequencies and initial phases of each replica creates a destructive interference between replicas so that the center of their distribution gradually stops moving.

This destructive interference (dephasing) depends on the individual features of the phonon subsystem. The higher the *phonon* density of states, the faster the dephasing occurs. Because the phonon density of states scales linearly with QD size, the crossover time τ is shorter for larger dots and longer for smaller dots. According to this consideration, the phonon bottleneck is reflected in the character rather then the relaxation rates. In the regime $t < \tau$, Eq. 2.15 demonstrates memory effects in small QDs, notably non-exponential carrier relaxation. In the regime $t > \tau$, Eq. 2.15 demonstrates the exponential form of carrier relaxation and no memory effects. The larger the QD, the shorter the memory effects are. For example, $\tau = 0.3$ ps for holes and $\tau = 0.5$ ps for electrons in the $Pb_{68}Se_{68}$, slightly shorter times compared to the calculated 0.4 ps for holes and 50.6 ps for electrons in the smaller $Pb_{16}Se_{16}$ QD. It is important to note that our approach allows only rough evaluation of the τ which indicates that the dephasing time is shorter than the estimated crossover time in QDs. For more accurate calculations of phonon-induced dephasing times in QDs, the TDKS molecular dynamics coupled with the framework of the optical response function and semiclassical formalisms were applied to $Pb_{16}Se_{16}$ and $Pb_{68}Se_{68}$ QDs and predicted dephasing times of around 10 fs [Kamisaka *et al.* (2006)].

For all investigated photo-excitation energies in the QDs we study, the relaxation occurs from 1.4 to 0.5 ps, clearly demonstrating the absence of the bottleneck even in such small clusters. Still, even the fastest relaxation is at least twice as slow as the CM time with an observed upper bound of 250 fs [Schaller and Klimov (2004); Schaller *et al.* (2006); Ellingson *et al.* (2005); Schaller *et al.* (2005a)]. Both electron and hole relaxation rates speed up with increased excitation energy. There is almost no rate dependence on the initial excitation for hole relaxation in CdSe due to the very dense VB in this type of QD. In both QDs, holes are faster than electrons. This speed is rationalized by a more dense VB compared to CB, as well as stronger NA coupling between hole states than electronic states. However, this difference is more sharply pronounced in CdSe which has a more asymmetric band structure; holes relax almost twice as fast as electrons. For $2.5 \times E_g$ and $3 \times E_g$ excitations, both electrons and holes in the CdSe QD relax slightly faster than carriers in the PbSe QD. This is in agreement with experimental observations of these QDs [Schaller *et al.* (2005d)]. In real systems, such a trend is even more exaggerated due to the very efficient Auger recombination process in CdSe which takes place through electron coupling with very dense hole states, and depressed coupling in PbSe from the nearly symmetric band structure.

2.4 Conclusions

We have simulated the phonon-mediated electron and hole dynamics in PbSe and CdSe QDs by the DFT-based TSH method in real time and at the atomistic level of detail. Our studies were partially motivated by the following mystery: the theoretical prediction of the electron-phonon relaxation bottleneck is challenged by experimental results detecting fast subpicosecond carrier relaxation in small QDs where the energy spacing is much larger than the energy of a single phonon. To explain these conter-intuitive results, three possible relaxation channels in QDs have been suggested: (1) through surface states, (2) through Auger recombination, and (3) through a multi-phonon process. We regard our numerical results to be in support of explanations 1 and 3.

Specifically, the dense distribution of electronic levels near the energy gap is observed due to surface reconstructions and the lack of absolute spherical symmetry in the QD surface. Ultimately, the proportion of these states to the core states is higher for smaller dots. Most of these states are optically dark and are not activated in absorption spectra, but they do couple to phonons and facilitate the relaxation process. In contrast to the common view that quantum confinement results in strong quantization of electronic energy levels in QDs, we find that the spacing between the state energies nearly matches the phonon frequencies except for a few levels near the energy gap. While the calculated optical absorption spectra of QDs indeed show discrete bands as observed experimentally, phonons of various symmetries couple most of nearly dark states. This coupling originates from surface reconstructions and generates an auxiliary pathway for carrier relaxation.

On the other hand, localization of wavefunctions and strong nonadiabatic electron-phonon coupling (numerically estimated as a few meV) in small QDs make a multi-phonon process highly probable. Our simulations demonstrate that elementary transition energies span a wide energy range, from 0.001 to 0.6 eV for different carriers and materials. This range includes carrier transitions far from resonance of any active phonon mode. These results support the hypothesis of a thermally-activated multi-phonon relaxation channel. Our simulations demonstrate that the phonon bottleneck is avoided through two relaxation pathways: surface states and multi-phonon processes, both of which are enhanced in smaller dots. Our results also reveal a stronger coupling of carriers with acoustic rather than optical phononsin both CdSe and PbSe QDs. The coupling with low-frequency acoustic phonons rationalizes the ultrafast dephasing time and large homogeneous line width of the optical transitions detected experimentally.

Additional results include (1) relaxation dependence on the QD material, (2) rise of carrier relaxation rates with incident photon frequency, and (3) numerical evidence of phonon memory effects. (1) PbSe and CdSe QDs exhibiting rocksalt and wurzite lattice symmetry are believed to have qualitatively different electronic band structures: a nearly symmetric CB and VB in PbSe and a strongly asymmetric CB and VB in CdSe QDs. There is still much debate about whether the PbSe QD truly exhibits mirror-like symmetry in its bands. Our simulations support the nearly symmetric structure of the DOS in PbSe QDs and the asymmetric DOS in CdSe QDs. Due to this strong asymmetry between the VB and CB of the CdSe QD (caused by denser levels in CB than VB), the hole relaxation occurs faster in CdSe compared to PbSe. Electron relaxation has almost the same rates for both tyes of QDs and demonstrates a multi-phonon character.

(2) We find that the higher the energy of the initial excitation, the faster the electron relaxation occurs in both materials. This is supported by (a) the increase in averaged nonadiabatic coupling with increased excitation energy; and (b) much higher density of levels for carrier states far from the gap than for carrier states close to the gap.

(3) Our simulations also constantly demonstrate an interesting feature of carrier relaxation: a transition from a Gaussian-like form to an exponential. The duration of the Gaussian regime of relaxation is expected to grow inversely proportional to the dot diameter as a partial "reincarnation" of the phonon bottleneck effect. Since simple rate models in the Markovian

limit typically generate exponential damping, we suggest that the simulated complex form of the carrier relaxation behavior can be attributed to phonon coherence. We also suggest that further focus on this issue may reveal the importance of non-Markovian memory effects.

The details provided by the reported simulations generate valuable insights into QD properties, reconcile the seemingly contradictory observations of wide optical line spacing and no phonon bottleneck to the relaxation, and rationalize why highly efficient carrier multiplication is possible in PbSe and CdSe nanocrystals despite the absence of the phonon bottleneck.

Chapter 3

Phonon-Induced Free Carrier Dynamics in Carbon Nanotubes

This chapter focuses on phonon-assisted excited charge carrier relaxation in semiconducting SWCNTs. Relaxation through the valence and conduction bands are addressed in analogy to the relaxation in CdSe and PbSe quantum dots that was considered in Chapter 2. Further, nonradiative recombination of cooled electrons and holes is investigated. Firstly, we will describe the experiments that motivated our studies. This includes a review of time resolved ultrafast spectroscopic experminents, which probe the intraband relaxation dynamics, and photoluminescence spectroscopy that studies the radiative recombination of electrons and holes across the band gap. A brief note on the electronic structure of SWCNTs that are relevant to the simulations will conclude our review. Secondly, the theory and methodology that compose our models will be discussed. We will refer the reader to previous chapters on the TDKS-FSSH scheme, which makes up the backbone of our model. Instead, the methodology will focus on a semi-classical model of dephasing in CNTs, the incorporation of decoherence on FSSH and simulation details specific to our study. Pure dephasing provides a measure of the coupling of the electronic and phonon subsystems and is responsible for the linewidths observed experimentally. Further, dephasing is a quantum phenomena and allows us to introduce quantum effects into the otherwise classical model of surface hopping. Thirdly, we present the results of our calculations of relaxation dynamics in semiconducting carbon nanotubes. Phonon-induced intraband relaxation in the zig-zag (7,0) CNT is detailed. Next estimates of the dephasing timescales in the (7,0), (6,4) and (8,4) CNTs and their corresponding fluorescence linewidths are explained. In order to further understand ambigious experimental results, we have studied two common defects in CNTs and modelled temperature effects on dephasing. Our final investigation is into the nonradiative recombination rates in the (6,4) CNT, with and without defects and at low temperature. We show that including decoherence effects into the FSSH model is pivotal for understanding the decay dynamics in CNTs and identify that defects may be responsible for the multiple decay timescales reported in the literature. Finally, we present a summary of our results and describe future challenges to our models and potential applications.

3.0.1 *Dynamics in Nanotubes: Optical Experiments*

Owing to recent improvement in the processing and production of SWCNTs, applications in molecular and quantum information technology are moving forward [Avouris and Chen (2006)]. The same as for QD-based devices, electron-phonon interaction [Shreve *et al.* (2007b)] and phonon-induced electronic dephasing [Roche *et al.* (2005a)] in SWCNTs play key roles in their technological applications. For instance, the response times of logic gates [Mason *et al.* (2004)], optical switches [Chen *et al.* (2002)] and lasers [Set *et al.* (2004b)] based on SWCNTs depend on the electron-phonon coupling. Energy loss that determines the conductivity of SWCNT wires [Terabe *et al.* (2005); Tans *et al.* (1997)], and field-effect transistors [Misewich *et al.* (2003); Maroto *et al.* (2007)] occurs by charge-phonon scattering. Electron-phonon interactions provide a mechanism for the observed superconductivity of SWCNTs [Tang *et al.* (2001); Bohnen *et al.* (2004a)] and create distortions in tube geometric structure [Tretiak *et al.* (2007b); Kilina and Tretiak (2007)]. Quantum dephasing due to the electron-phonon coupling sets limits on coherent spin [Hueso *et al.* (2007)] and charge [Jarillo-Herrero *et al.* (2006)] transport through SWCNTs. Clearly, insight into the charge-phonon interaction, relaxation, and dephasing mechanisms in these nanomaterials is extremely important.

During the last decade, numerous optical experiments have been conducted to obtain data on excitation dynamics in carbon nanotubes. While, Raman and Raleigh scattering experiments have deeply probed the electronic structure and electron-phonon interactions in CNTs, time-resolved (ultrafast) optical experiments allow one to obtain data on the dynamics of carriers in these systems [Dresselhaus *et al.* (2007)]. The amount of progress made in under a decade towards understanding the nature of optical excitations is quite impressive. A large amount of the credit should go to synthetic developments, as well as the separation and isolation of single carbon nanotubes (see Section 1). Initial ultrafast spectroscopic measurements were performed on bundles of CNTs, also known as 'bucky paper'.

These ensembles contained a mixture of metallic and semiconducting CNTs, with varying diameters, and a mix of carbonaceous impurities. The early experiments interpreted the ultrafast response of CNTs as a two timescale mechanism [Hertel and Moos (2000)]. A fast timescale, on the order of hundreds of femtoseconds, was assigned to electron-electron scattering and relaxation of the nonequilibrium charge carrier distribution. The slower timescale, on the order of tens of picoseconds, was attributed to phonon-induced relaxation of the electrons to the lattice temperature [Hertel and Moos (2000)].

Improvements in separation and purification techniques [O'Connell *et al.* (2002); Hertel *et al.* (2006)] removed the nonradiative decay channels provided by the metallic tubes and tube bundles in the bucky paper allowing the study of luminescence in SWCNTs and opened the door to many new questions. The increasing interest in and availability of CNT samples produced an upsurge in the number of groups studying them experimentally. CNTs were studied as they fluorescence from micelles in solution [O'Connell *et al.* (2002)], from polymer matrices [Hagen *et al.* (2005)], and even when grown across silicon pillars [Lefebvre *et al.* (2003)]. These diverse studies have led to a large variation in the reported interband and intraband relaxation timescales, as well as fluorescence quantum yields, showing the strong dependence of experimental data on sample preparation, excitation energy, light intensity, and detection techniques.

Reported timescales for relaxation from the second excited band (E_{22}) to the band edge (E_{11}) ranged from tens to hundreds of femtoseconds [Htoon *et al.* (2005); Manzoni *et al.* (2005a); Lanzani *et al.* (2005)] to over a picosecond [Hertel and Moos (2000); Huang *et al.* (2004); Korovyanko *et al.* (2004b); Zamkov *et al.* (2005); Ostojic *et al.* (2004); Wang *et al.* (2004)]. Eventually, the relaxation times were found to depend strongly on the intensity of the light source [Ma *et al.* (2005c, 2004)]. High laser intensities were shown to create multiple excitons that would undergo exciton-exciton annihilation or Auger processes. In an Auger process, two excitons diffuse together and one of the electron-hole pairs recombines giving its energy to the other electron-hole pair. This causes the electron and hole to move to higher energy states in the E_{11} subband, which then decay via interactions with phonons.

It is now generally recognized that relaxation from the E_{22} band occurs in less that 100 fs at low excitation intensity. It was also shown that relaxation rates in nanotubes strongly depend on tube morphology and tube environment. Thus, the relaxation is slower in isolated SWCNTs than in SWCNT bundles and SWCNT-polymer mixtures, where tube-tube and tube-environment interactions provide additional decay channels [Hertel and Moos (2000); Zamkov *et al.* (2005)]. The relaxation is faster in larger SWCNTs, since they have more dense states, which also causes a more rapid decay of higher energy excitations [Hertel and Moos (2000); Zamkov *et al.* (2005)].

Measurements of the fluorescence lifetimes are as scattered as the initial intraband relaxation timescales. Reported decay rates range from tens to hundreds of picoseconds, and fitting functions show single, bi- and multi-exponential timescales [Jones *et al.* (2005); Hagen *et al.* (2005); Seferyan *et al.* (2006); Jones *et al.* (2007); Berger *et al.* (2007)]. Theoretical calculations of the radiative lifetime in CNTs predict hundreds of picoseconds to several nanoseconds, while experimental estimates vary from tens to hundreds of nanoseconds. It has been speculated that nonradiative decay channels due to carbonaceous impurities (i.e. fullerene-type fragments or small very defective CNTs), and small bundles of semiconducting and metallic SWCNTs in the samples may explain some of the uncertainty in the measurements. In one study, three (6,4) SWCNTs embedded in a polymer matrix were identified and their fluorescence timescales measured. The reported decay rates were 24, 60 and 183 ps for the same chirality tube, indicating the large effect of the local environment [Hagen *et al.* (2005)].

Also the fluorescence spectra from individual nanotubes with identical structures demonstrate different emission energies and linewidths that likely arise from the defects or environment perturbations. Thus, suspended SWCNTs show a fluorescence linewidth of 10-12 meV [Lefebvre *et al.* (2004a)], while solution or polymer based samples have linewidths of 20-25 meV [O'Connell *et al.* (2002); Hartschuh *et al.* (2003b)]. This suggests a broadening of approximately 10-15 meV due to interactions with the environment. The optical linewidth detected for fluorescence or stimulated emission is determined by either the lifetime of the transient or the phonon-induced dephasing. If the lifetime is much greater than the dephasing time, dephasing determines the linewidth [Mukamel (1995)]. Dephasing of optical excitations is a quantum effect controlled by the interactions of the electronic system and the nuclei. Motions of the ions cause the electronic wavefunction to begin spreading amongst other other states. The less interaction or coupling between the electronic state and the nuclei, the longer the wavefunction is allowed to spread. This is, in a sense, a manifestation of the uncertainty principle [Skinner (1988)]. Eventually, interactions with the ionic subsystem cause the wavefunction to collapse back to its initial state (provided it did not radiatively decay to a lower energy state first). Detected optical linewidths provide a measurement of the spreading of the electron, and, hence, the dephasing time. Dephasing may be detected directly using photon echo spectroscopy, but this has not yet appeared in the literature.

Further complicating matters is a surprisingly low fluorescenc quantum yield, with initial reports ranging from 10^{-4} [Wang *et al.* (2004); Hagen *et al.* (2005)] to 10^{-3} [Jones *et al.* (2005)]. A more recent report, using CNTs grown across silicon pillars, has raised the figure to 0.07 [Lefebvre *et al.* (2006)], suggesting that extra nonradiative decay channels in solution or polymer matrices may have a signifcant role in the dynamics of SWCNT samples. Unlike most other molecules studied to date, the fluorescence intensity or spectrum from a single nanotube unexpectedly did not fluctuate. This absence of spectral and intensity fluctuations is in stark contrast to fluorescence from individual semiconductor quantum dots and from most molecules, which exhibit an emission intermittency or on-off blinking behavior for all excitation intensities on time scales that span many orders of magnitude [Dresselhaus *et al.* (2007)]. However, at higher excitation powers, SWCNT fluorescence intensity fluctuations can be observed that are likely due to laser-induced sample heating or the interaction of multiple excitons. The observation that SWCNTs show no emission intensity blinking or bleaching demonstrates that SWCNTs have the potential to provide a stable, single-molecule infrared photon source with extremely narrow linewidth [Avouris and Chen (2006)].

Another important aspect of SWCNTs are possible defects in the hexagonal carbon lattice. Lattice defects occur in all real materials and SWCNTs are no exception. Possible defects in SWCNTs include monovacancies, monomer and dimer insertions, bond rotations, and impurity additions [Sternberg *et al.* (2006)]. Defects can have large effects on the mechanical and electronic properties of bulk and nanostructures systems and are often the site of chemical interactions. Fortuitously, SWCNTs are relatively defect free, and one investigation into high quality samples showed defect densities of approximately 1 in 10^{12} atoms [Fan *et al.* (2005)]. This corresponds to about one defect every four 4 microns in a SWCNT. Still, even at very low concentrations, defects can disrupt the conjugated π systems of CNTs and effect transport behaviour and other important characteristics of CNTs [Roche *et al.* (2005b)]. In addition, recent development of novel hybrid materials such as nanotubes functionalized by polymers, DNA [Zheng *et al.* (2003b,a)], or dyes [Casey *et al.* (2008, in press)], intercalated [Nasibulin *et al.* (2007); Yanagi *et al.* (2007)] and merged [Krasheninnikov and Banhart (2007)] tubes calls for more rigorous consideration of the defect problem in SWCNTS, especially in the interface and merged areas where chemical perturbation or irradiations introduce local disorder on a tube surface.

In light of the mentioned complications, complexities, and variety of experimental data a description of the electronic structure and carrier dynamics in SWCNTs at the atomistic level would help elucidate many of the obscured details.

3.0.2 *Electronic Structure*

The electronic structure of nanomaterials and its changes upon photoexcitation effect many important characteristics of nanoscale optoelectronic devices, including carrier transport and luminescence efficiency. Therefore, a better understanding of electronic and optical properties of nanosystems reveals fundamental physical phenomena and has important technological implications. Experimentally, the electronic structure of semiconductor SWCNTs is mainly probed by means of absorption, fluorescence, and Raman spectroscopies [Weisman and Bachilo (2003); Bachilo *et al.* (2002); Fantini *et al.* (2004b); Telg *et al.* (2004); Doorn *et al.* (2004)]. Early optical spectra in SWNTs were interpreted in terms of free electron-hole carriers. As predicted by tight-binding Hamiltonian models, these single-particle states are grouped into equally spaced subbands of valence and conduction bands with diverging density of states at band-edges known as van Hove singularities (vHS) [Saito *et al.* (1998a)], labeled by $\pm1, \pm2, \ldots$, *etc.*, with increasing energies. The transitions arising among these manifolds are typically labeled as E_{ij}, where indices i and j refer to the valence and conduction band singularities, respectively, as schematically presented in Fig. 3.1.

In general, vHS were observed in many low-dimensional systems [Hugle and Egger (2002)]. The DOS for a d-dimensional system with dispersion relation $E(\mathbf{k})$ can be written as an integral over the Fermi surface, $\upsilon(E) = 2\pi^{-d} \int dS/|\nabla_k E(\mathbf{k})|$. The quantity in the denominator is the group velocity. Due to symmetries in a crystal, the group velocity may vanish at certain momenta, resulting in a divergent integrand. This divergence is integrable in three dimensions, and typically leads to a finite DOS. In lower dimensions, however, very pronounced vHS appear. In SWCNTs, considered as the 1-D limit, the vHS diverges like $\upsilon(E) \sim 1/\sqrt{|E - E_n|}$, when the

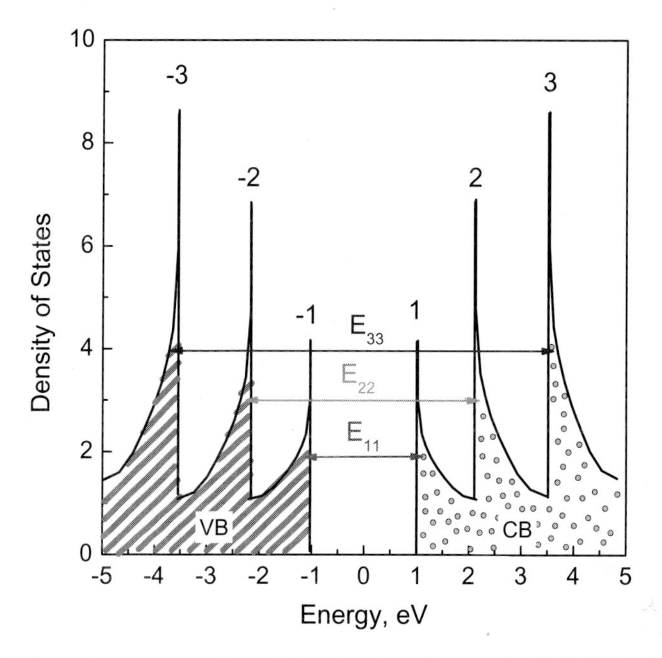

Fig. 3.1 Schematic representation of the density of states in SWCNTs. Van Hove singularities are marked by $\pm1, \pm2, \pm3\ldots$. The optical transitions are defined by arrows and marked by E_{ii}.

energy E approaches the threshold E_n [Hugle and Egger (2002)]. Therefore vHS appear as sharp features in the energy-dependent DOS reflecting the onset of new active sub-bands. However, the ratio between excitonic energies violates predictions of conventional one-electron theories because of electronic correlation effects [Kane and Mele (2004)].

Recent studies established that the quasi-one-dimensional structure of the SWCNTs and a small dielectric constant lead to high electron-hole binding energies and the formation of strongly-bound excitons (composite electron-hole pairs) as the primary photoexcited species. Theoretical studies [Kane and Mele (2004); Spataru *et al.* (2004); Zhao and Mazumdar (2004a)] followed by transient spectroscopy and nonlinear absorption experiments [Ma *et al.* (2005b); Korovyanko *et al.* (2004a); Gambetta *et al.* (2006b)] have unambiguously revealed that the photophysics of SWNTs is dominated by excitons with typical binding energies of 0.2-0.5 eV [Wang *et al.* (2005a); Maultzsch *et al.* (2005); Perebeinos *et al.* (2004); Chang *et al.* (2004); Tretiak *et al.* (2007b); Kilina and Tretiak (2007); Tretiak (2007a)], depending on the tube diameter and chirality. Electronic correlation effects give rise to several excitonic bands associated with each transition between van Hove peaks. Each resulting manifold contains both optically allowed (bright) and optically inactive (dark) exciton states, as well as a continuum band of free carriers [Zhao *et al.* (2006); Wang *et al.* (2006b)].

Excitons are covered in much more detail in the next chapter, while here we focus on continuum states of free carriers in SWCNTs. Below it will be shown that TDKS qualitatively predicts the correct interaction between the ions and the electron and hole wavefunctions. This allows for comparison to experiment and to draw conclusions on the nonradiative decay channels in SWCNTs.

3.1 Theoretical Considerations

3.1.1 *Semiclassical Model of Dephasing*

The theory of the lineshape of optical absorption and emission has been extensively studied and is quite advanced [Skinner (1988); Mukamel (1995)]. The observed linewidth can be decomposed into an inhomogeneous term, which accounts for the distribution of optically active species and their different local environments, and a homogeneous term that is fundamental to the species of interest. Neglecting the inhomogeneous broadening, the linewidth Γ is determined by the dephasing time, T_2, composed of the excited-state lifetime T_1 and the pure-dephasing time T_2^*:

$$\Gamma = \frac{1}{T_2} = \frac{1}{T_1} + \frac{1}{T_2^*}. \tag{3.1}$$

Pure-dephasing is associated with fluctuations in the electronic energy levels. The fluctuations occur due to interactions between the electronic subsystem and other subsystems, such as phonons or solvent. If the excited-state lifetime is much longer than the pure-dephasing time, Γ is determined by the pure-dephasing T_2^*. This limit is assumed for the rest of the derivation and will be justified later. The dephasing function may be computed directly using the flucations of the energy for the electronic transition. In this direct method, the dephasing function is defined as [Mukamel (1995)]

$$D(t) = \exp(i\omega t)\left\langle \exp\left\{-\frac{i}{\hbar}\int_0^t \Delta E(\tau)\, d\tau\right\}\right\rangle_T . \tag{3.2}$$

Here, ω is the thermally averaged energy gap divided by $\hbar$ and ΔE is the energy gap of interest. The brackets denote thermal averaging over a canonical ensemble of initial conditions.

To further investigate the pure-dephasing, we calculate the dephasing times using a cumulant expansion [Mukamel (1995)]. Here, we first compute the autocorrelation function (ACF) of the electronic excitation energy:

$$C(t) = \langle \Delta E(t) \Delta E(0) \rangle_T. \tag{3.3}$$

The ACF is unnormalized and the brackets again denote thermal averaging over a canonical ensemble of initial conditions. Oscillation and decay of the ACF characterize the electron-phonon interaction memory. The initial values of the ACF give the average fluctuation of the excitation energy due to coupling to phonons.

In this case, the ACF (3.3) is then doubly integrated

$$g(t) = \int_0^t d\tau_1 \int_0^{\tau_1} C(\tau_2)\, d\tau_2. \tag{3.4}$$

The function $g(t)$ is then exponentiated to define the approximate dephasing function

$$D(t) = e^{-g(t)}. \tag{3.5}$$

The approximate dephasing function agrees with the direct function (3.2) if the higher order cumulants beyond the second order ACF are insignificant. In general, the higher order cumulants disappear when the motions of the ions remain in the harmonic limit.

3.1.2 *Decoherence Effects*

By treating nuclei classically, the original FSSH scheme [Tully (1990); Parahdekar and Tully (2005)] excludes coherence loss that occurs in the electronic subsystem by coupling to quantum vibrations. Decoherence can be neglected if it is slower than the electronic transition time, for instance, in the sub-picosecond relaxation of the $E_2 2$ band in SWCNTs. The decay to the ground state occurs much more slowly, and decoherence effects must be explicitly included in the quantum-classical simulation. We implemented the loss of quantum coherence within TDDFT-FSSH using a simple semiclassical approach [Schwartz *et al.* (1996); Prezhdo and Rossky (1997a)], which was tested extensively with a variety of condensed phase physical [Kamisaka *et al.* (2006)], chemical [Hwang and Rossky (2004); Larsen and Schwartz (2006)] and biological [Lockwood *et al.* (2001); Kim and Hammes-Schiffer (2006)] systems. The expansion coefficients of the electronic wave-function in the Kohn-Sham basis (2.12) were allowed to evolve coherently up to the decoherence time, at which point they are reset to 0 or 1 with the probabilities given by the squares of coefficients (2.13). In the absense of decoherence, the non-adiabatic mixing of the electronic states can occur to a much greater extent, creating significantly larger transition probabilities. Loss of coherence decreases the transition rate, as exemplified in by the quantum-Zeno effect [Prezhdo and Rossky (1998); Streed *et al.* (2006); Maniscalco *et al.* (2006)]. in the fast decoherence limit.

3.1.3 *Simulation Details*

The SWCNT studies focus on the smallest available semiconducting CNTs, which are shown in (Fig. 3.2). The (7,0) SWCNT is achiral and has a small unit cell of 28 atoms. The simulation cell used 4 unit cells for a total of 112 atoms to increase the phonon spectrum [Bohnen *et al.* (2004b)]. Chiral SWCNTs have much larger unit cells and the simulation cells were only able to incorporate a single cell. The (6,4) cell has 152 atoms, while the (8,4) cell has 112 atoms. These are two of the smallest chiral SWCNTs and the number of atoms in the unit cell quickly increases to hundreds or thousands of atoms for large diameter SWCNTs. The chiral (6,4) and (8,4) SWCNTs are among the smallest tubes accessible experimentally [O'Connell *et al.* (2002); Hagen *et al.* (2005)]. The zigzag (7,0) has not been identified experimentally, but was initially chosen to minimize the size of the electronic basis and simulation cell. The structure of all the SWCNT are generated using TubeGen 3.3 [Frey and Doren (2005)], with a C-C bond length of 1.42 Å. The simulation cell was constructed using periodic boundary conditions along the axis of the SWCNT and 8 Åof vacuum in the direction perpendicular to the axis to prevent spurious interaction. The SWCNT geometry and cell size are relaxed to the minimum energy structure.

In order to further investigate the effects of disorder in the lattice, two common defects [Sternberg *et al.* (2006)] were introduced into the (6,4) CNT, (Fig. 3.2). The Stone-Wales (SW) defect is a rotation of a C-C bond within the ideal hexagonal lattice. The 7557 defect represents the insertion of an extra C-C dimer into the lattice. Both C_2 defects disrupt π-conjugation and create two five-membered rings and two seven-membered rings. The geometries of the defect sites are different in the two cases though, with the dimer insertion creating a stronger distortion. The unit cell of the (6,4) tube easily accommodates the defects parallel to the tube axis, which is the lowest energy configuration [Sternberg *et al.* (2006)], and allows for proper geometric relaxation of the defects sites.

The simulations are performed in VASP [Kresse and Furthmüller

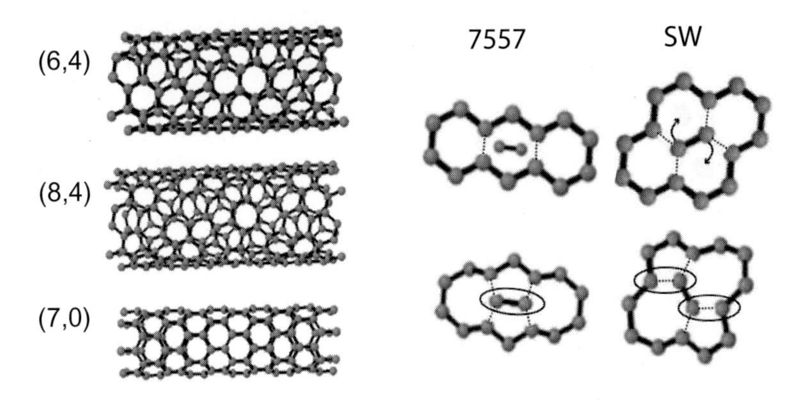

Fig. 3.2 Simulation cells of the SWCNTs (left) and defects (right) used in the study. The (7,0) tube is achiral and has small unit cell of 28 atoms. Four units cells are used in the simulation cell. The (6,4) and (8,4) tubes are chiral and each simulation cell consists of only one unit cell.

(1996a)]. The Perdew-Wang generalized gradient approximation (GGA) [Perdew (1991)], Vanderbilt ultrasoft pseudopotentials [Vanderbilt (1990)], and converged plane-wave basis sets are employed. As customary with pure DFT functionals, the excitation energy was adjusted to the experimental value [Hagen *et al.* (2005)] using the scissor operator [Pulci *et al.* (1998)]. The same adjustment was applied to the (6,4) CNT with defects. The simulation predicts that the lowest energy exciton is weakly allowed, in agreement with the experiments [Wang *et al.* (2005b); Seferyan *et al.* (2006); Berger *et al.* (2007); Mortimer and Nicholas (2007); Metzger *et al.* (2007); Zaric *et al.* (2006)] and other electronic structure calculations [Spataru *et al.* (2005a); Perebeinos *et al.* (2005c); Zhao and Mazumdar (2004b); Tretiak *et al.* (2007a); Dresselhaus *et al.* (2007)]. As SWCNTs are direct band gap semiconductors, only the Γ point was sampled in these studies.

After heating the system to 300 K (50 K) by repeated velocity rescaling, a microcanonical trajectory is run in the ground electronic state with a 1 fs time step. The microcanonical trajectory of 1.5 ps was used for the intraband relaxation and a trajectory of up to 4.0 ps was used in the dephasing and nonradiative recombination simulations. The intraband relaxation was modelled using the traditional FSSH-TDKS scheme, while the nonradiative recombination was simulated with the FSSH-TDKS scheme including decoherence. In both studies, 400 initial conditions are sampled for the relaxation dynamics, as well as 10,000 surface hopping trajectories. For the intraband relaxation, the states within the singularities are chosen based on the strongest transition dipole moment at a given initial time. Over a third of all excitations occur between two pairs of electron and hole states. Only the HOMO and LUMO states were used for the nonradiative recombination simulations, modelling the situation where hot carriers relax to the band edge and thermalize before recombining.

3.2 Results and Discussion

3.2.1 *Phonon-Induced Intraband Relaxation*

The electron and hole relaxation under investigation is initiated by an excitation from the van Hove singularities below the Fermi level to the vHS above the Fermi level, as in ultrafast laser experiments [Ma *et al.* (2004); Manzoni *et al.* (2005a); Korovyanko *et al.* (2004b)]. The vHs dominate the

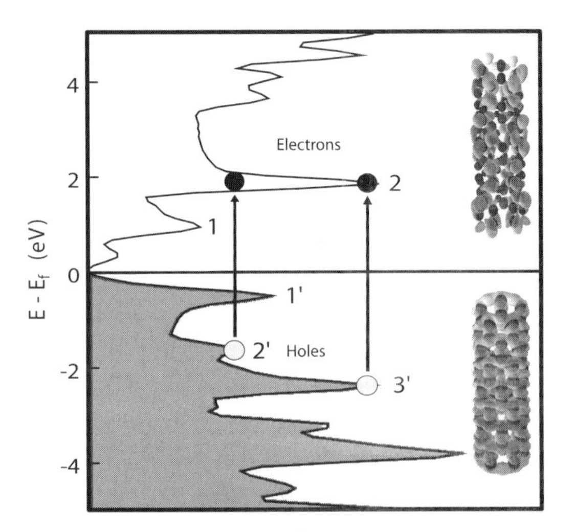

Fig. 3.3 Schematic illustrating the modeled ultrafast spectroscopic experiments. Electrons are excited from the vHs in valence band to vHs in the conduction band, leaving holes below the Fermi energy. The electrons and holes then relax to the corresponding band edge.

electronic spectrum of the nanotubes due to their large oscillator strengths and can be used to identify the chirality of the tube. The vHS arise due to the quantization of electronic states around the CNT axis [Saito *et al.*

(1998b)], (see Fig.3.3). As discussed in the introduction, curvature, strain and electron-hole correlations cause an asymmetric distribution of states in the valence and conduction bands. Fig.3.3 shows that the holes have a larger densities of state than the electrons.

After the initial excitation, the electrons and holes relax nonadiabatically through the DOS to their respective band edges. The relaxation of the total electron and hole energies are presented in Fig. 3.4. The electrons and

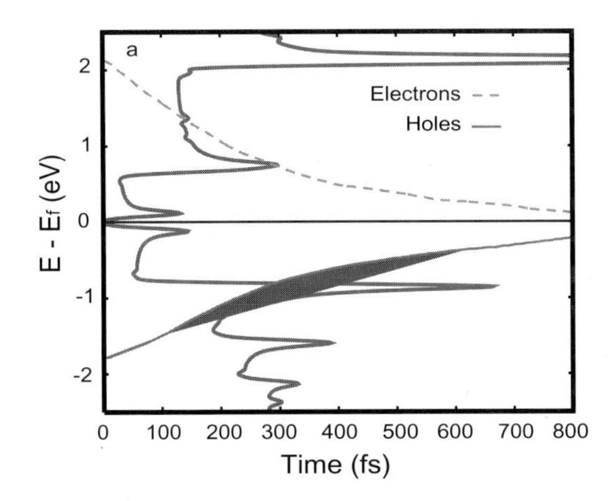

Fig. 3.4 The dynamics of electron and hole relaxation through the DOS of the (7,0) SWCNT are depicted. The electron decay is purely exponential, while the hole decay is slower and has both a Gaussian and exponential component.

holes show different dynamics in relaxing through the VB and CB. That the holes decay more slowly than the electrons is quite surprising, since the larger DOS of the holes should facilitate faster relaxation. A Gaussian and an exponential component are clearly visible in the hole relaxation (Fig. 3.4), while a Gaussian component can hardly be distinguished in the electron relaxation. The energy decay is fit with the sum of the Gaussian and exponential components

$$\Big|E - E_f\Big|(t) = \Big|E - E_f\Big|(0) \quad \Big[Ae^{-t/\tau_e} + (1-A)e^{-(t/\tau_g)^2}\Big] \qquad (3.6)$$

with the fitting parameters listed in Table 1. The exponential component dominates the electron decay. Holes show both Gaussian and exponential decay. The amplitude of the Gaussian component, which is often associated with coherent quantum dynamics [Prezhdo and Rossky (1998); Prezhdo (2000); Luis (2003); Exner (2005)], is restricted by the lower density of electronic states, generating larger energy gaps between the states that are coupled by phonons. and favors incoherent inelastic electron-phonon scattering. The larger and broader DOS below the Fermi energy allows the hole to spread and reside in the second singularity longer than the electron. At 800 fs almost 20% of holes still reside within the second vHS. This number is less than 5% for the electrons.

The two-component hole decay is promoted by strong coupling of holes to both high and lower frequency phonons. In contrast, electrons interact more strongly with the high frequency modes. The spectra densities of the energies of the most optically active electron and hole state pairs shown in Fig. 3.5 were computed from the microcanonical trajectory that was used to sample the photoexcited states for the NA dynamics. The G-type longitudinal optical (LO) phonons with frequencies around 1500 cm^{-1} provide the fastest relaxation pathway for both electrons and holes. This LO phonon also corresponds to C-C stretch frequencies. The stronger coupling of the holes to the radial breathing mode (RBM) and other acoustic modes at frequencies below 500 cm^{-1} can be rationalized by the better match between the nodal structures of the hole states and RBMs. Partial charge densities

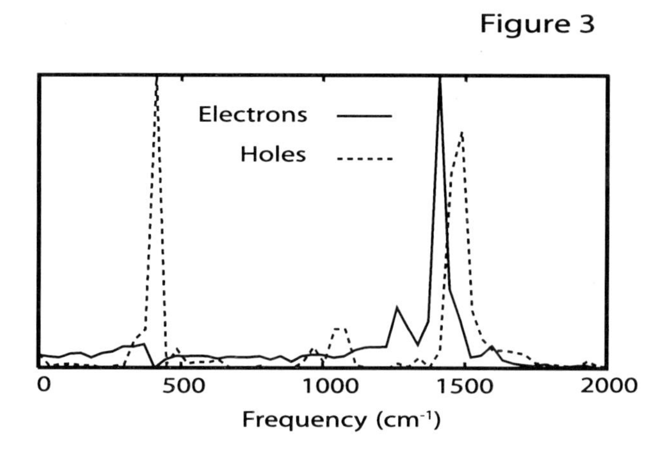

Fig. 3.5 Spectral density created by Fourier transforming the energies of important initial electron and hole states. As can be seen, electrons couple more strongly to optical phonons near 1500 cm^{-1}, while holes couple to both acoustic and optical phonons.

of the electron and hole states are shown as insets in (3.3). The lower energy VB states that support holes have fewer nodes than the conduction band states that support electrons, and RBMs have fewer nodes than LO phonons. As a result, RBMs couple better to the VB states. The stronger coupling to the lower frequency RBMs explains why the hole relaxation is slower than the electron relaxation, even though holes have a higher DOS.

The timescales for the electron and hole relaxation computed for the (7,0) tube are within the range of in initial reports of ultrafast spectroscopic experiments [Htoon *et al.* (2005); Manzoni *et al.* (2005a); Ma *et al.* (2004); Hertel and Moos (2000); Huang *et al.* (2004); Korovyanko *et al.* (2004b); Zamkov *et al.* (2005); Lanzani *et al.* (2005); Ostojic *et al.* (2004); Wang *et al.* (2004); Ma *et al.* (2005c)]. The calculations support the picture where the photoexcited state scatters into the continuum of hot electron and hole pairs, which relax by phonon emission [Lanzani *et al.* (2005); Ma *et al.* (2005a)]. Once the carriers relax to the bandedge, localized excitons are formed. The involvement of RBM in the CNT charge carrier dynamics seen in our simulation has been detected by both current/voltage [LeRoy *et al.* (2004a)] and spectroscopic [Htoon *et al.* (2005)] measurements. The coupling of the electronic system to the LO phonons has been detected in the Raman spectra [Oron-Carl *et al.* (2005)]. The reported *ab initio* NA dynamics also agree with the tight-binding electronic structure calculations, in which the electron-phonon scattering was dominated by the RBM and LO phonons in two narrow frequency regions [Perebinos *et al.* (2005a,b)].

3.2.2 *Estimates of Dephasing Timescales in SWCNTs*

As mentioned in previous sections, if the excited-state lifetime is much longer than the pure-dephasing time, Γ is determined by the pure-dephasing. Recent experiments [Jones *et al.* (2005); Zhang *et al.* (2005)] and theoretical calculations [Spataru *et al.* (2005b); Perebeinos *et al.* (2005c)] on fluorescence of single-walled CNTs has produced excited-state lifetime expectancies on the order of nanoseconds. As dephasing timescales are expected to be on the order of tens to hundreds of femtoseconds, this assumption seems valid. Dephasing and optical linewidths are determined by interactions of the electronic subsystem with the nuclei and environment. As discussed in the introduction, environmental factors can be very important in the linewidth and nonradiative decay channels. Due to computational complexities and system size, in our present calculation pure-dephasing occurs solely via interaction of the electronic subsystem with CNT phonons. ACFs for the lowest electronic excitations of the ideal (6,4) tube at 300 K and 50 K, and the (6,4) tube with the 7557 and SW defects at 300K were calculated from the microcanonical trajectory and are shown in (Fig. 3.6).

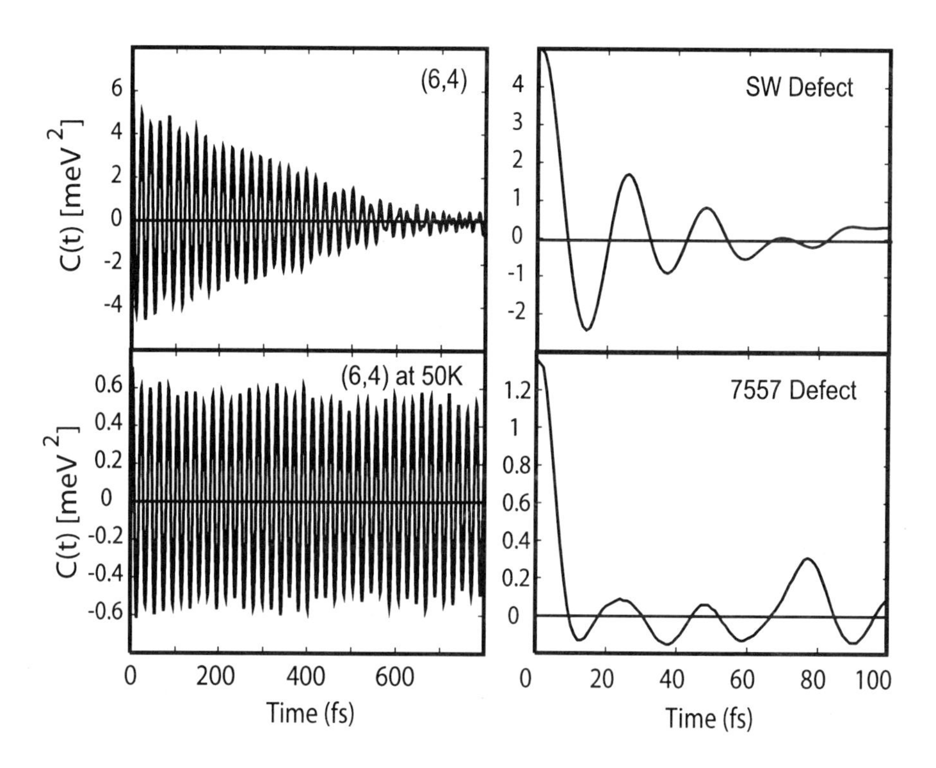

Fig. 3.6 The calculated ACFs of the energy gaps for the lowest energy excitation are shown. Note the difference in timescales (x-axis) between the ideal (left) and defected (right) tubes.

Note that the left and right sections of Fig. 3 have significantly different time scales and that all panels have different y-axis scales. The value of the ACF at $t = 0$ gives the average fluctuation of the excitation energy due to coupling to phonons. The ACFs decay much more slowly in the ideal CNTs than in the those with the defects. Comparing the ideal tubes at the ambient and low temperatures we observe that the ACF at 300 K decays in just under a picosecond, while the ACF at 50 K has decayed little in the same amount of time. Further, the energy fluctuation is also an order of magnitude larger at 300 K. Contrastingly, in the CNTs with defects, the ACFs decay very rapidly, within tens of femtoseconds. This is due to the geometric disorder, which induces a wider range of vibrational modes in these tubes. The 7557 defect induces a faster decay of the ACF than the SW defect, and the 7557 ACF is less symmetric with respect to the time axis. This is attributed to the larger distortion of the CNT geometry created by the insertion of a C_2 dimer. The C_2 dimer creates a bulge in the sidewall of the CNT, while a C-C bond rotation has a much lesser effect. The amplitudes of the energy fluctuation form the following sequence for the (6,4) tubes at 300K in decreasing order: ideal, SW defect, and 7557 defect. This sequence correlates with the localization of the electronic excitation and the number of carbon atoms that are able to couple to the electronic subsystem. Similar calculations were done for the (8,4) and (7,0) SWCNTs as explained above. The (8,4) tube was very similar to the (6,4), while strain in the (7,0) SWCNT caused a faster decay of the ACF. Data was not included for brevity, however the analysis that follows includes these two other nanotubes.

The CNT phonon modes that contribute to the decay of the ACF(3.6) and that are responsible for the pure-dephasing are analyzed in (Fig. 3.7). The figure presents the Fourier transforms (FT) of the ACFs, known as the

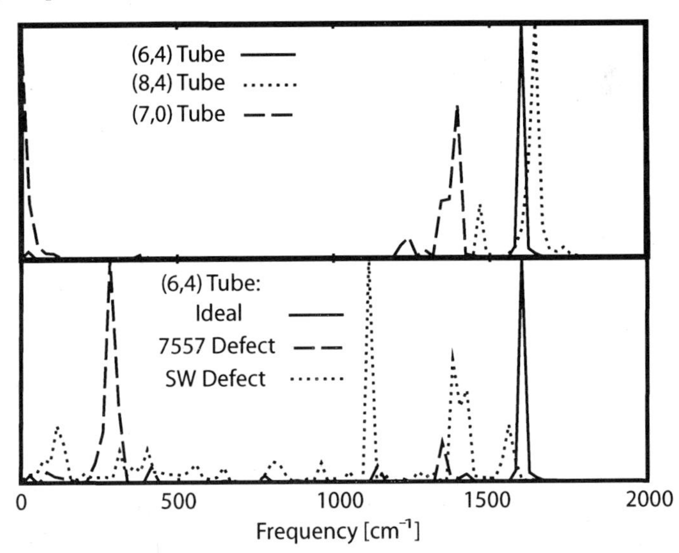

Fig. 3.7 Schematic of the spectral densities of the ideal and defected tubes studied. The (6,4) SWCNT has the same spectral density at 50K and 300K. The disorder induced by the defects promotes coupling to a wide range of vibrations.

influence spectrum or spectral density [Skinner (1988)]. The upper panel presents the data for the three ideal tubes, and the lower panels compares the ideal (6,4) tube with those perturbed by the defects. All FTs shown are based on the 300 K simulations, as the ideal (6,4) tube at 50K exhibits the same frequencies as the (6,4) tube at 300 K. The electronic excitation energies of the (6,4) and (8,4) CNTs fluctuate primarily due to coupling to the optical G-mode around 1600 cm^{-1}. The frequency of the mode is slightly higher in the (8,4) tube due to less geometric strain and stronger chemical bonding as the (8,4) tube has a larger diameter. The G-mode of the (7,0) tube is significantly lower due to the substantial geometric strain.

As shown in the bottom panel of Fig. 3.7, defects introduce a much broader range of vibrations. Stronger coupling to acoustic modes, the radial breathing mode (RBM) and disorder modes is evident in the spectrum of the tubes with the defects. The coupling to a larger range of frequencies rationalizes the dramatically faster decay of the ACF of the tubes with defects compared to the ideal tubes, (Fig. 3.6). The SW defect couples to a very wide range of frequencies, while the 7557 defect primarily couples to a local mode in the RBM frequency range. This is due to the fundamental differences between the two defects, even though they result in similar bonding patterns, (Fig. 3.2). The SW defect is rotation of a bond, which introduces a great deal of disorder in the phonon spectrum of the whole tube. The 7557 defect, however, creates a local vibrational mode. Since the excitation density is strongly localized on the added C-C bond, it is the vibration of this local structure that induces the electronic dephasing.

Now that the ACFs have been addressed and the phonon modes that are important to dephasing have been identified, we proceed to calculating the dephasing functions. The direct dephasing functions for all tubes, calculated using Eq. 3.2 are shown in Fig. 3.8. The top panel depicts the results for the ideal SWCNTs at 300 K, while the lower panel presents the data for the ideal (6,4) tube at 50 K and 300 K, and compares the ideal tube with the defects. The cumulant expansion based dephasing functions are not shown here, though the validity of the second order approximation can be evaluated based on the fits reported in Tables 1 and 2. The approximation is valid in all tubes considered here except for the (7,0) tube, which exhibits long memory effects in the electron-phonon interaction, most likely due to its small diameter and strong geometric strain. The cumulant function (3.5) well agrees with the direct function (3.2) up to the first oscillation occuring in the latter at around 10fs. At later times, the cumulant function smoothly continues the rapid decay initiated within the first 10fs, while the direct function changes its angle and decays more slowly, similarly to the dephasing functions for the other ideal tubes.

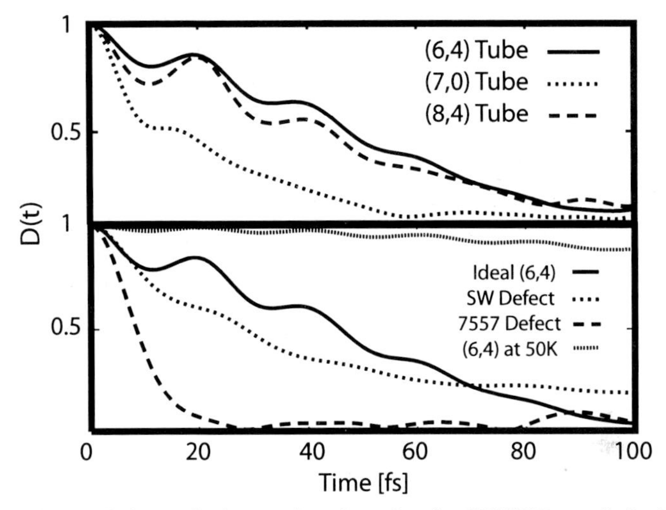

Fig. 3.8 Calculated direct dephasing functions for the SWCNTs studied. The (7,0) tube's dephasing function decays more rapidly than the other ideal tubes due to its narrow diameter and large amount of strain. The defects cause the dephasing to increase significantly, while at 50 K the dephasing takes much longer.

The direct and cumulant dephasing functions were fit by the equation

$$f(t) = \exp\left(\frac{-t}{\tau}\right)\frac{1 + A\cos(\omega t)}{1 + A}. \tag{3.7}$$

The fitting parameters are shown in (Tables 3.1 and 3.2), for the direct (3.2) and cumulant (3.5) dephasing functions, respectively. The frequency ω represents the oscillation of the dephasing function, and the constant A is related to the amplitude of the oscillation. The time τ is the pure-dephasing time. For the ideal tubes and the (6,4) tube with the SW defect, $\omega \approx 0.30$ fs^{-1} corresponds to the optical G-mode. The 7557 defect has $\omega = 0.12$ fs^{-1} originating from the local mode in the RBM frequency range. The oscillation of the dephasing functions for the 7557 defect is rather weak.

The calculated dephasing times of (6,4) and (8,4) tubes in the ideal geometry are between 50 and 60fs and correspond to linewidths of 11-13meV, Eq. (3.1). These linewidths are within the range of 10-15meV reported by Lefebvre *et al.* [Lefebvre *et al.* (2004a)] and are slightly more narrow than those reported by O'Connell *et al.* [O'Connell *et al.* (2002)]. The former reference investigated nanotubes grown across silicon pillars, which are not able to interact strongly with an outside environment. The broader peaks seen in the latter work can be attributed to solvent or surfactant interac-

Pure-dephasing time T_2^* and corresponding homogeneous linewidths Γ, Eq. (3.1), obtained by fitting the direct dephasing functions (3.2) shown in Fig. 5 to Eq. (3.7).

Tube	T_2^* (fs)	A	ω (fs^{-1})	Γ (meV)
(8,4)	51.2	0.061	0.30	12.8
(7,0)	23.2	0.10	0.29	28.2
(6,4)	59.6	0.086	0.32	11.0
7557	10.2	0	–	64.6
SW	48.0	0.062	0.25	13.7
50K	955	0.0056	0.30	0.69

Pure-dephasing times T_2^* and corresponding homogeneous linewidths Γ, Eq. (3.1), obtained by fitting the cumulant dephasing function (3.5) to Eq. (3.7).

Tube	T_2^* (fs)	A	ω (fs^{-1})	Γ (meV)
(8,4)	52.4	0.062	0.30	12.5
(7,0)	9.12	0	–	72
(6,4)	54.3	0.080	0.30	12.1
7557	10.1	0	–	65.2
SW	42.1	0.027	0.24	15.6
50K	750	0.0045	0.30	0.87

tions with the nanotubes, as the experiment were done in solution. Since our simulations excluded solvent, our results should be compared directly with Ref. [Lefebvre *et al.* (2004a)], indicating very good agreement with the experiment. The (7,0) tube has a dephasing time of approximately half that of the other ideal geometry tubes, Table 1. The faster dephasing is attributed to the larger fluctuation of the excitation energy and coupling to a broader range of phonon modes, (Fig. 3.7). The (7,0) linewidth still is within the experimental range.

The two defects have strong, but different effects on the dephasing rate. The 7557 defect increases the dephasing time by a factor of six. This is due to strong electronic coupling to a local vibrational mode created by the defect. The coupling is notably anharmonic, as indicated by the asymmetric form of the ACF, (Fig. 3.6). The SW defect only slightly accelerates the dephasing. Even though the electronic excitation localized on the SW defect couples to a wider range of phonons, the coupling is harmonic and relatively weak due to the delocalized nature of the SW defect state.

By comparing the initial values of the ACFs (Fig. 3.6), one can see that the magnitude of the fluctuation of the excitation energy is larger for the SW defect than for the 7557 defect. Initially, one might have expected faster dephasing in the SW case. This is not true, however, because the ACF of the SW state is quite symmetric, generating a significant amount of cancelation in the integral (3.4) and slowing the decay time. The dephasing time of the (6,4) tube at 50 K is decreased by over an order of magnitude from room temperature. This is attributed to smaller motions of the ions due to lower thermal energy. The suppression of ionic motion, in turn, creates less of a perturbation to mix the electronic states and results in a slow dephasing time. Our dephasing time and the corresponding linewidth is in good agreement with the reported ultrathin linewidths at 50 K [Htoon *et al.* (2004)].

3.2.3 *Nonradiative Recombination in a Semiconducting SWCNT*

Assuming laser excitation energies that are less than the ionization threshold, the relaxation of excited carriers to the ground state may occur radiatively or nonradiatively. In radiative decay, energy from the charge carriers is released as a photon as the electron and hole recombine. In nonradiative recombination, the excess energy is transfer to the lattice in the form of phonons. These two process may occur on similar timescales and can be competitive. The focus of this work is to estimate the nonradiative lifetime τ_r of excited electrons and holes. Others have done estimates of the radiative lifetime [Spataru *et al.* (2005b); Perebeinos *et al.* (2005c)] by carefully calculating the energy levels and transition dipole moments. The reader is referred to these works for deeper analysis.

The estimates of the dephasing time in SWCNTs developed previously allow one to modify the FSSH scheme to include quantum effects on the electronic evolution. As FSSH is a semi-classical theory, a lack of decoherence can have a great effect on relaxation timescales. We employ the FSSH with decoherence to study the nonradiative recombination in a semiconducting SWCNT and investigate the importance of dephasing on the nonradiative decay. Figure 3.9 shows the fluctuation of the first excitation energy induced by the nuclear motion. The average fluctuation in the ideal (6,4) tube is around 0.1 eV at room temperature and decreases approximately a factor of five when the temperature is lowered to 50 K. Defects increase the energy fluctuation by a factor of two, with the 7557

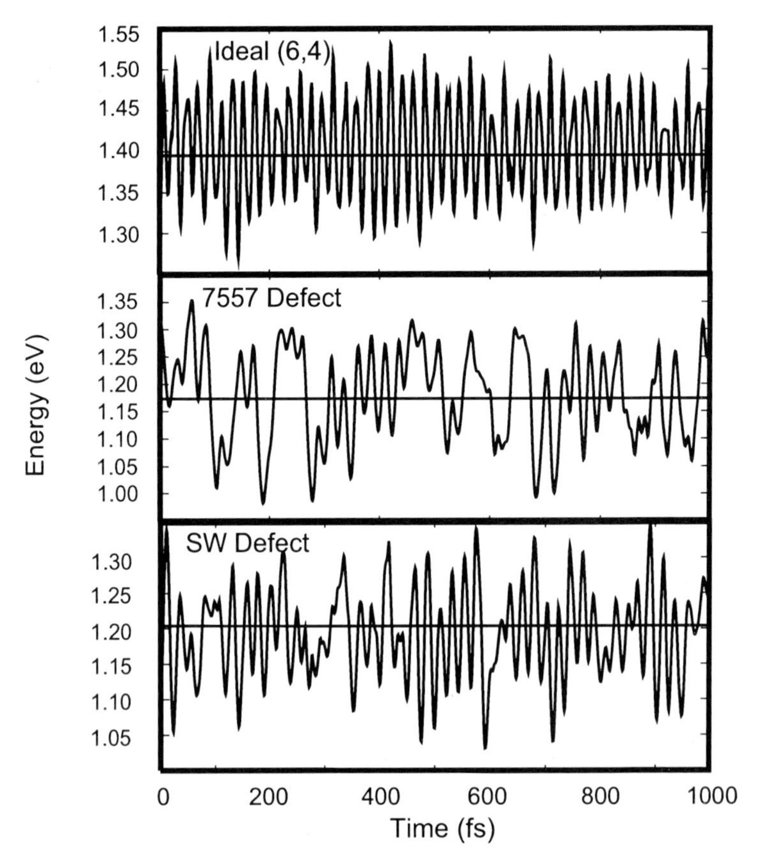

Fig. 3.9 Fluctuations of the First Excitation Energy.

defect producing a bigger change than the SW defect. The increase in the average fluctuations is because defects generate stronger electron-phonon coupling by localizing vibrational modes [Sternberg *et al.* (2006)] and electronic states. Defects further lower the energy of the first excitation by creating states in the bandgap. The average excitation energy for the ideal (6,4) SWCNT is 1.39 eV, while the 7557 and SW defects reduce the gap energy to 1.17 eV and 1.20 eV, respectively. The average excitation energy of the ideal tube increases slightly to 1.46 eV with decreasing temperature due to thermal expansion of the CNT, in agreement with experiments [Lefebvre *et al.* (2004b); Metzger *et al.* (2007)].

The electron-phonon coupling is generated in the ideal (6,4) CNT by the high frequency G-modes, (Fig. 3.7), both at high and at low temperatures. The 50 K spectral density is the same as that at 300K. The SW defect shows many disorder modes over a broad range of frequencies [Vandescuren *et al.* (2007)]. Many of the disorder modes couple to the electronic transition. In contrast, the excitation localized on the 7557 defect couples primarily to a single low-frequency mode, (Fig. 3.7). The insertion of a C-C dimer notably distorts the CNT geometry [Sternberg *et al.* (2006)], creating a local mode in the range of low frequency radial breathing modes.

Fig. 3.10 and Table 3.3 characterize the nonradiative decay of the ideal (6,4) CNT at low and ambient temperatures, as well as the decay rates calculated for the two defective SWCNTs. A linear growth of the ground electronic state population during the first 3ps following the relaxation to the bandedge is predicted. Assuming that the full nonradiative decay is exponential, as observed in the experiments [Jones *et al.* (2005); Hagen *et al.* (2005); Seferyan *et al.* (2006); Jones *et al.* (2007); Berger *et al.* (2007); Wang *et al.* (2004); Ma *et al.* (2006); Hirori *et al.* (2006)], we fitted the linear 3 ps component determined by the simulation to $P(t) = 1 - \exp(t/\tau) \approx t/\tau$ and obtained the nonradiative recombination times τ reported in Table 3.3. The relaxation time shows strong dependence on the pure-dephasing time, which is given in the first row. The second row provides the relaxation time using the true dephasing times [Habenicht *et al.* (2007)], while the last two rows illustrate the strong effect of quantum decoherence on the relaxation process.

The nonradiative lifetime estimated for the ideal (6,4) tube at room temperature is around 150ps, falling within the experimental range [Jones *et al.* (2005); Hagen *et al.* (2005); Seferyan *et al.* (2006); Jones *et al.* (2007); Berger *et al.* (2007); Wang *et al.* (2004); Ma *et al.* (2006); Hirori *et al.* (2006)]. The defects substantially accelerate the relaxation, reducing the

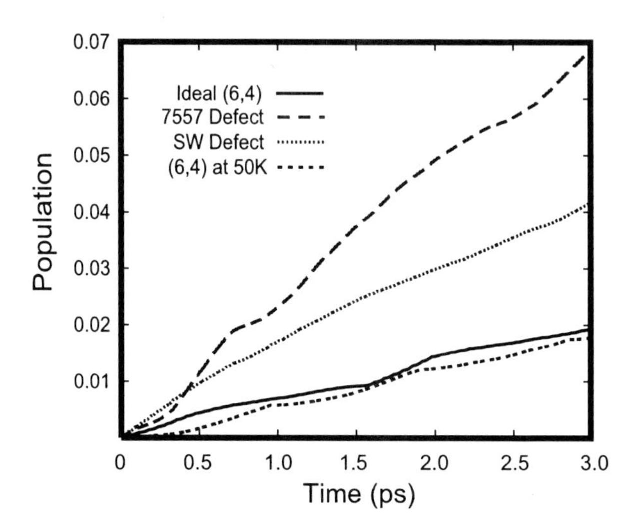

Fig. 3.10 Nonradiative decay of the first excitation is shown for the tubes studied. The defects substantially increase the decay, however, the decay shows little temperature dependence.

Tube	Est. Dephasing (fs)	Relax Time (ps)	10 fs Dephasing Relax Time (ps)	500fs Dephasing Relax Time (ps)
Ideal	59.6	147.3	716.2	39.8
7557	18.4	41.5	77.5	3.55
SW	48.0	66.7	251.3	16.2
50K (6,4)	955	170	4,906	258

decay time to tens of picoseconds and providing a rationalization for the multiple decay components observed in the experiments. For example, Refs. [Hagen *et al.* (2005); Berger *et al.* (2007)] report two sets of decay times around 10-20 ps and 180-300 ps. Our calculations indicate that the longer times correspond to pristine tubes, while the shorter times are due to defects. The bond insertion, 7557 defect, has a more profound effect on the relaxation process than does the bond rotation SW defect.

The influence of defects on the nonradiative decay rate can be traced to the following three factors. First, defects lower the excitation energy, (Fig. 3.9), thereby better matching the the electronic energy to the phonon frequency and increasing the decay rate. Second, they create strongly localized electronic states and vibrational modes, (Figs. 3.7) generating larger electron-phonon coupling. Third, defects accelarate dephasing, (Table 3.1). The first two factors act to speed up the relaxation by creating a better electron-phonon energy match and a larger coupling. The third factor slows the relaxation in general, as exemplified by the quantum Zeno effect [Prezhdo and Rossky (1998); Streed *et al.* (2006); Maniscalco *et al.* (2006)], although counter-examples are known [Prezhdo (2000)]. Overall, the defects accelerate the relaxation, because the reduced excitation energy and, particularly, the stronger NA coupling dominate over the increased dephasing rate. Thi is consistent with the recently observed quenching of mobile excitons at localized sites [Cognet *et al.* (2007)].

The nonradiative decay shows little temperature dependence, (Fig. 3.10) and (Table 3.3). This is not surprising, since the relaxation is induced by the modes whose frequencies are significantly higher than k_BT at the relevant temperatures [Ma *et al.* (2006)]. The fact that we were able to obtain weak temperature dependence semiclassically supports our simulation approach and emphasizes the importance of including decoherence into the quantum-classical simulation [Schwartz *et al.* (1996); Prezhdo and Rossky (1997a)]. The decay rate is determined by the product of the electron-phonon coupling matrix element squared and the correlation function for the phonon motions associated with the ground and excited electronic states, as in the time-domain version of Fermi's Golden rule. Both the squared coupling and the correlation function depend on the phonon kinetic energy, but in the opposite ways. In the semiclassical description [Schwartz *et al.* (1996); Prezhdo and Rossky (1997a)], the squared coupling increases with temperature, but the correlation function also decays faster, resulting in the cancellation.

The moderate temperature dependence of the nonradiative decay rate seen in our calculations agree well with the excited state lifetime experiments [Metzger *et al.* (2007); Ma *et al.* (2006); Hirori *et al.* (2006); Karaiskaj *et al.* (2007)]. Our result also agrees with the PL intensity data reported in Ref. [Lefebvre *et al.* (2004b)], although other authors [Berger *et al.* (2007)] observe much stronger dependence of PL intensity on temperature. According to our calculations, the weak temperature dependence of the PL intensity is due to the weak temperature dependence of nonradiative decay, rather than insignificant nonradiative decay channels, as thought originally [Lefebvre *et al.* (2004b)]. Note that in contrast to the PL decay, the PL linewidth dramatically decreases at low temperatures [Htoon *et al.* (2004)] and is determined by the pure-dephasing time rather than by the excited state lifetime [Habenicht *et al.* (2007)].

Our simulation results fall within the range of the experimental data [Jones *et al.* (2005); Hagen *et al.* (2005); Seferyan *et al.* (2006); Jones *et al.* (2007); Berger *et al.* (2007); Wang *et al.* (2004); Ma *et al.* (2006); Hirori *et al.* (2006)], still, a number of approximations may have influenced the estimated decay times. Likely most important for the simulation, the relaxation rate shows very strong dependence on the pure-dephasing time, (Table 3.3). Although the estimated pure-dephasing times are in good agreement with the experimental linewidths, the semiclassical scheme used to incorporate the dephasing effect into the TDDFT simulation was very simple. It should be emphasized that this is the first time dephasing was included in time-domain DFT and that more sophisticated approaches are significantly more computationally expensive. It was also assumed, due to computational complexity, that the decay bottleneck is determined by the transition from the lowest singlet excited state to the ground state. Since the first exciton creates a sub-band of states, some of which are thermally accessible from the lowest energy state [Mortimer and Nicholas (2007)], the decay can occur from several states simultaneously. Moreover, the singlet manifold co-exists with the lower-energy triplet manifold [Tretiak (2007b)], and the excitation can intersystem-cross into triplets due to spin-orbit coupling. Triplets are typically longer lived than singlets and may account for the nanosecond components in the experimental data.

3.3 Conclusions

We have used TDDFT with the FSSH scheme to investigate free carrier-phonon dynamics in SWCNTs. TDDFT was used to model the electron-phonon dynamics, while the FSSH scheme was employed for the electron relaxation and electron-phonon backreaction. Our initial studies into the relaxation of hot electrons and holes in the (7,0) zig-zag SWCNT found good agreement with experimental results. By validating our approach, we were able to further interpret the data and atomistic dynamics. We were able to identify a multiple timescale relaxation component for holes and a single exponential relaxation for electrons. The multiple timescale relaxation was attributed to a higher DOS of hole states that promoted equilibration of the hot carriers manifest as a strong Gaussian component of the relaxation. Once thermalized within the VB singularity, the hole decayed to the bandedge on an exponential timescale. This Gaussian component is also promoted by the stronger coupling of VB states to the RBM. The larger spacing of the CB levels and the coupling of CB states primarily to high energy LO phonons prevented such a thermalization of the electrons, and resulted in a purely exponential decay. It should be noted here that larger (longer) simulation cells of the same SWCNT could increase the decay timescales. States within the vHs occur because of quantization of electronic states *around* the SWCNT axis. The states between the vHs (i.e. between the E_{22} and E_{11} vHs) depend on wavefunctions confined *along* the length of the SWCNTs. Thus, longer simulation cells should provide more intermediate states, leading to smaller hopping energies between states in both the VB and CB. Regardless, the generalizations of our results, especially the role of the G-mode and RBM will be uneffected by large simulations cells.

In our next study, we were able to calculate the timescales of phonon-induced dephasing in semiconducting SWCNTs. The dephasing timescales are directly proportional to the optically observed absorption and emission linewidths determined by photoluminescence experiments. The optical linewidths are an extremely fundamental property in spectroscopy as they directly reflect the influence of the nuclei on the energetic localization of the electronic wavefunction. Our results are in excellent agreement with the reported emission linewidths of SWCNTs suspended across silicon pillars and in vacuum. This allows us to predict that interactions with the environment, such as solvent, surfractant and substrate, broaden the emission peaks by a factor of two. Further, we can postulate that defects substantially increase the dephasing rate and may be identified by wider linewidths. This increase in the dephasing rate is caused by the lattice distortion introduced by the defects and, subsequentally, stronger coupling to a wider range of vibrational modes. Finally, we were able to show that dephasing is strongly temperature dependent and results in sub-meV linewidths at 50K, also in agreement with experimental observations.

Our investigations into the dephasing timescales further allowed us to probe the nonradiative relaxation in SWCNTs. The TDDFT-FSSH model with decoherence allowed us to incorporate the quantum effect of dephasing in a semi-classical description of the electron-phonon interaction. Using the calculated dephasing times, we were able to reproduce experimental timescales of photoluminescent decay, and show that defects may be responsible for shorter decay times, as well as multi-exponential decay profiles observed experimentally. Further, our results indicate that the weak temperature dependence of the photoluminescence is due to the cancellation of two competing factors. The nonadiabatic couplings are smaller at lower temperature because of smaller motions of the ions. This works to decrease the nonradiative decay rate. Second, the decoherence timescales are much longer at lower temperatures. This allows more time for the electronic wavefunctions to spread among the states, increasing the nonradiative decay rate. We find that these two factors largely cancel, resulting in little temperature dependence.

In summary, we have used a semi-classical model to simulate the phonon-induced intraband relaxation, dephasing and nonradiative recombination in semiconducting SWCNTs. The effects of defects and temperature on the lattice vibrations and coupling to the electronic subsystem was also explored. The introduction of a simple dephasing model allowed us introduce quantum effects to the electron-phonon interactions and elucidate its importance. This investigation has provided important insight into the fundamental dynamics of SWCNTs and the interpretation of experimental results. It is also the first *ab initio* simulation of the phonon-induced relaxation and dephasing in SWCNTs.

Chapter 4

Including Electron-hole Correlations: Excitonic and Vibrational Properties of Carbon Nanotubes

In this chapter we present quantum-chemical studies of excited state electronic structures of finite size semiconducting SWCNTs using ESMD methodology previously successfully applied to describe conjugated polymers and other organic molecular materials. The results of our calculations were originally published in Refs [Shreve *et al.* (2007a); Araujo *et al.* (2007); Tretiak *et al.* (2007b); Kilina and Tretiak (2007); Kilina *et al.* (2008)]. The beginning of this chapter presents brief review of experimental works demonstrating the importance of excitonic and vibrational effects in SWCNTs. Next, we overview the exciting theoretical approaches allowing consideration of electron-hole and electron-phonon interactions. Most of these theories focus either on excitonic or vibrational phenomena; both of these effects have to be taken into account for the proper description of excited-state molecular dynamics in SWCNTs. Our ESMD methodology allows this consideration. The extended description of the ESMD is presented in Section 4.2. Based on this approach we investigate coupled excitonic and vibrational effects in SWCNTs for a number of different tubes emphasizing emerging size-scaling laws. The results of our simulations are presented in Section 4.3; they show quantitative agreement with available spectroscopic data.

First we investigate the diameter dependence of energy gaps and energies of first three optically allowed excitons in SWCNT. Next we analyze the photoinduced changes in charge densities and bond-orders. This provides an understanding of localization/delocalization properties of excitons in SWCNTs. Spontaneous uneven distribution of the π electrons over the bonds (i.e., Peierls dimerization) is observed throughout the entire nanotube when the system is in the ground state, particularly in large-radius SWCNTs. Vibrational relaxation following photoexcitations leads to some local distortions of the tube surface overriding the Peierls dimerization. Such a decrease in surface corrugations affects the electronic system and forms localized states – self-trapped excitons. We also present the results of calculations of exciton-phonon couplings: associated Huang-Rhys factors, Stokes shift, and vibrational relaxation energies; they all increase with enlargement of tube diameters. The simulated exciton-vibrational phenomena would be possible to detect experimentally, allowing for better understanding of photoinduced electronic dynamics in nanotube materials.

4.1 Introduction

4.1.1 *Experimental Evidence of Excitons and Strong Exciton-Phonon Coupling in SWCNTs*

New synthetic advances and applications of sophisticated experimental approaches have resulted in a revolutionary re-evaluation of the basic photophysics of SWCNTs over the past five years. Formally, optical transitions in SWCNTs are associated with electronic transitions E_{ii} between i-th van Hove singularity at the VB and i-th van Hove singularity at the CB [Saito *et al.* (1998a)] (see Fig. 3.1). However, the ratio between excitonic energies violates predictions of conventional one-electron theories because of electronic correlation effects [Kane and Mele (2004)]. Here we use the notation E_{ii} adopted from one-electron models, while considering the excited states in SWCNTs as excitons with a strong electron-hole Coulomb coupling.

In contrast to QDs, recent transient spectroscopy and nonlinear absorption data [Ma *et al.* (2005b); Korovyanko *et al.* (2004a); Manzoni *et al.* (2005b)] have unambiguously revealed that photophysics of SWCNTs is dominated by strongly bound excitons (interacting electron-hole pairs), rather than free particles. Measurements of the excitonic binding energy were done using two-photon excitation spectroscopy [Wang *et al.* (2005a); Maultzsch *et al.* (2005)] and pump-probe spectroscopy [Zhao *et al.* (2006)]. Taking advantage of allowed and well defined transitions to both the bound exciton state and to the near-continuum unbound states, binding energies of ~ 0.4 eV for the first excitonic band (E_{11}) in semiconducting SWCNTs with 0.8 nm diameters were found. This is about one hundred times larger than that for bulk semiconductors, but comparable to other 1-D materials, such as conjugated polymers [Scholes and Rumbles (2006)]. Measurement of the binding energy of the second exciton E_{22} by resonant Raman scattering provides even higher values of ~ 0.5 eV and ~ 0.6 eV for (10,3) and

(7,5) tubes, respectively [Wang *et al.* (2006a)]. Starting from the pioneering study by Ando [Ando (1997)], a large amount of theoretical work has confirmed strong excitonic effects in SWCNTs [Kane and Mele (2004); Spataru *et al.* (2004); Zhao and Mazumdar (2004a); Perebeinos *et al.* (2004); Chang *et al.* (2004)]. These results demonstrate that, unlike in corresponding bulk systems and QDs, the excited-state properties of nanotubes are dominated by many-body interactions. This is a typical scenario for many 1-D materials [Scholes and Rumbles (2006)]. Thus, excitonic effects can not be neglected or treated as a small perturbation to the band gap of SWCNTs, as was assumed for QDs.

Electron-hole correlations affect not only optically allowed (bright) transitions, but optically inactive (dark) states associated with each transition between van Hove peaks. For example, splitting of nearly doubly degenerate HOMO-LUMO transition E_{11} into 4 distinct excitonic transitions [Zhao and Mazumdar (2004a)] is schematically shown by red lines in Fig. 4.1 (a). This introduces a complex structure of overlapping inter-band states with

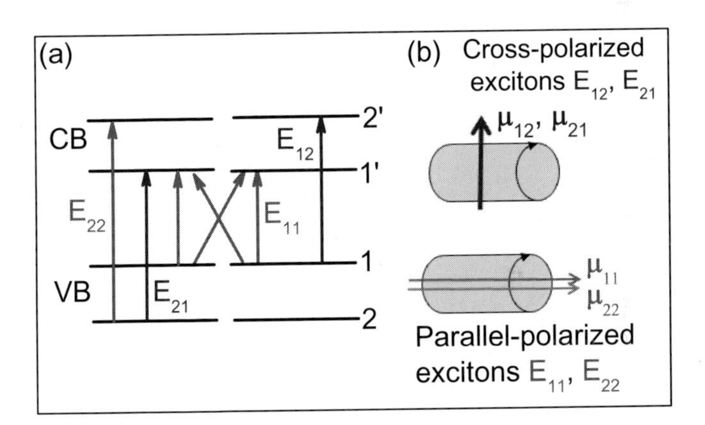

Fig. 4.1 Schematics of optical transitions in SWCNTs corresponding to collinear, parallel-polarized (red and green color) and perpendicular, cross-polarized (blue color) excitations (a) as illustrated by directions of the respective transition dipole moments μ_{ij} (b). For color reference, turn to page 154.

different angular momenta [Uryu and Ando (2006)], complicating dramatically the electronic structure of SWNTs and affecting their photophysical properties. For example, the energy position of the dark excitons with respect to the bright ones impacts the nanotube luminescence. Recent theoretical [Zhao and Mazumdar (2004a); Perebeinos *et al.* (2005c)] and experimental [Shaver *et al.* (2007)] studies explain the typical low photoluminescence quantum yield in SWNTs by the existence of dark excitons below the first optically bright exciton state, which trap much of the exciton population. The existing experiments have focused mostly on the optically allowed fundamental excitonic transitions (E_{11}, E_{22}, and so on), while the information on the entire excitonic spectra of SWNTs remains incomplete.

The dependence of optical response on the polarization direction of the incident light with respect to the tube axis can be utilized to extend investigations to other excitonic bands in SWNTs. For example, for light polarized parallel to the tube axis, only E_{ii} excitations are allowed. The corresponding strong E_{ii} peaks have been clearly observed in optical spectra and investigated intensively during the past few years. In contrast, the selection rules for perpendicular polarization (cross-polarization) of light [Gruneis *et al.* (2003)] allow optical absorption between subbands, where quasi-angular momenta differ by one, such as E_{12} and E_{21}. There are, however, very limited studies of such transitions. Within one-electron π theory, transverse optical absorption occurs at an energy that is exactly in the middle of the two lowest longitudinal absorption energies [Wang *et al.* (2007)]. Phenomenological implementation of electron-hole interaction and depolarization effects to the effective mass Hamiltonian shifts E_{12} and E_{21} energies to the blue; the shift amplifies with increase of Coulomb coupling matrix elements [Uryu and Ando (2006)]. Theoretically, it is predicted that the intensity of cross-polarized peaks is reduced drastically in SWNTs, due to the depolarization effects that screen the electric field through a self-consistent induced charge [AJIKI and ANDO (1994)]. Nonetheless, recent results of photoluminescence anisotropic spectroscopy report resolved absorption peaks for the polarization perpendicular to the SWNT axis, although the intensity of these peaks is much smaller compared to that of parallel-polarized transitions [Miyauchi *et al.* (2006); Lefebvre and Finnie (2007)]. Investigations of intermediate-frequency Raman modes [Fantini *et al.* (2005)] in bundled SWNTs also suggest the existence of cross-polarized excitations [Luo *et al.* (2007)]. However, the structure of cross-polarized excitons remains ambiguous, and calls for further detailed studies.

In addition to the excitonic features, carrier-phonon interaction plays an important role in SWCNTs. This made evident through several experimental findings. Among them are vibronic resonances in the photoluminescence excitation spectra [Chou *et al.* (2005); Htoon *et al.* (2004)], vibrational progression in absorption/fluorescence lineshapes [Qiu *et al.* (2005); Plentz *et al.* (2005a)], Raman spectroscopy [Goupalov *et al.* (2006); Fantini *et al.* (2004b); Telg *et al.* (2006)], coherent phonon excitation experiments [Gambetta *et al.* (2006b); Lim *et al.* (2006)], transport measurements [LeRoy *et al.* (2004b)], as well as theoretical studies [Habenicht *et al.* (2006); Perebeinos *et al.* (2005a); Figge *et al.* (2001); Tretiak *et al.* (2007b); Jiang *et al.* (2005); Machon *et al.* (2006)]. It is important to note that both electron-hole and electron-phonon interactions are typical features of 1-D systems [Scholes and Rumbles (2006)] such as conjugated organic [Heeger *et al.* (1988); Tretiak and Mukamel (2002); Wu *et al.* (2006)] and organometallic [Batista and Martin (2005)] polymers, mix-valence chains [Dexheimer *et al.* (2000)], and organic molecular materials [Klessinger and Michl (1995); Balzani and Scandola (1991); Herzberg (1950)]. However, nanotubes have a rigid structure and cannot be regarded as pure 1-D systems because of their circumferential dimension. Subsequently, excitonic and vibrational effects in SWCNTs are expected to differ from that in the above systems and to be very sensitive to tube size and geometry. Quantitative measurements of electron-phonon coupling constants and Huang-Rhys factors from the analysis of experimental Raman profiles [Yin *et al.* (2007); Shreve *et al.* (2007a)] and calculations [Shreve *et al.* (2007a); Machon *et al.* (2005)] place SWCNTs in a regime of weak coupling strength compared to typical molecular systems, but far exceeding electron-phonon coupling constants observed in semiconductor bulk materials.

Thus, both excitonic and vibronic effects are equally important for correct description of photoinduced dynamics in nanotubes. These phenomena need to be clearly understood to achieve proper functionalities of future nanotube-based electronic devices.

4.1.2 *Available Theoretical Approaches Calculating Excitonic and Vibrational Effects*

An accurate description of excitonic and vibrational features in SWCNTs depends on a proper incorporation of both strong Coulomb and exciton-phonon couplings. Computational studies utilizing accurate quantum-chemical methods are complicated and usually involve significant numerical effort. Consequently, existing theoretical investigations based on rigorous first principle methodologies focus on only one class of phenomena: either excitonic or vibrational. On the other hand, computations utilizing model Hamiltonian approaches have limited accuracy and do not include, for example, curvature-induced $\sigma-$ and $\pi-$ bond mixing and possible deviations from the sp^2 hybridization, which are important effects in the case of SWCNTs [Figge *et al.* (2001, 2002); Bohnen *et al.* (2004a); Dubay *et al.* (2002)].

One of the most accurate theoretical descriptions of excitonic effects in SWCNTs have been provided by a Green's functions approach via the solution of the Bethe-Salpeter equation (BSE) by adding self-energy corrections (GW) to the local-density approximation[Onida *et al.* (2002)]. This method has demonstrated significant electron-hole interactions and has proved the dominating role of excitons in the optical spectra of SWCNTs [Spataru *et al.* (2004); Chang *et al.* (2004)]. The same BSE approach but coupled to a much simpler empirical tight-binding Su-Schrieffer-Heeger (SSH) model has been fruitfully used to evaluate various excitonic properties [Perebeinos *et al.* (2005c)], size-scaling laws [Perebeinos *et al.* (2004)], carrier transport [Perebeinos *et al.* (2005b, 2006)], and electron-phonon coupling effects [Perebeinos *et al.* (2005a)] in a broad range of nanotube species. The DFT has also been widely used to calculate curvature, electron-phonon effects [Machon *et al.* (2006, 2005)], uncorrelated gaps [Barone *et al.* (2005a,b)], and non-adiabatic electron-phonon dynamics [Habenicht *et al.* (2006)]. Fewer applications of TDDFT to SWCNT systems have been reported. The latter technique allows for adequate treatment of excitonic effects with accuracy comparable to BSE-based methods [Casida (1995); Onida *et al.* (2002)] and has currently become a mainstream approach for calculating electronic excitations in molecular materials. For example, TDDFT application to narrow tubes [Reining *et al.* (2003)] demonstrates surprisingly good agreement with experimental results and reveals the importance of depolarization effects.

The above-described theoretical approaches assume periodic boundary conditions to address long SWCNT lengths. Recently, a number of simulations using finite tube lengths have been reported [Zhao and Mazumdar (2004a); Zhao *et al.* (2006); Zhou *et al.* (2004); Gambetta *et al.* (2006b); Tretiak *et al.* (2007b)]. These techniques use methodologies previously developed for molecular materials. Finite-size calculations are able to address all excitonic phenomena as once the molecular size becomes larger than the characteristic exciton size all physical properties become additive. This means, for instance, a saturation to constant excitation energies. Consequently, a large enough molecule well represents the infinite system limit. Such techniques are routinely applied to many other 1-D systems such as conjugated polymers or mixed-valence chains [Heeger *et al.* (1988); Tretiak and Mukamel (2002); Brédas *et al.* (1999); Dexheimer *et al.* (2000); Hutchison *et al.* (2003); Gierschner *et al.* (2007)]. This allows well tested, well understood, and accurate theoretical methodologies, developed for characterization of molecular systems, to SWCNTs.

For example, calculations based on empirical Pariser-Par-Pople (PPP) π-electron Hamiltonian [Zhao and Mazumdar (2004a)] explain the low PL efficiency of SWCNTs because of intrinsic low-lying 'dark' excitonic states. The same approach indicated many common features in the electronic spectra of SWCNTs and conjugated polymers; a fact that was confirmed by transient absorption spectroscopic measurements [Zhao *et al.* (2006)]. We further note that group-theoretical analysis of optical selection rules based on the periodic boundary conditions in SWCNTs fails to predict their strong nonlinear absorption, which should be completely forbidden in theory but clearly shows up in two-photon PL [Wang *et al.* (2005a); Maultzsch *et al.* (2005)] and transient absorption spectra [Zhao *et al.* (2006)]. In contrast, finite tube calculations are able to explain the basic photophysics underlying these spectra well and to characterize the properties of the excited states involved [Zhao *et al.* (2006); Tretiak (2007a)]. Consequently, solid state-like (periodic boundary conditions) and molecular-like (finite-size) theoretical approaches are complimentary in discovering the rich photophysics of SWCNT materials.

4.2 Computational ESMD Methodology

This work overviews an extensive study of both excitonic and vibrational effects in SWCNTs using finite-size molecular-type approaches. It is based on an excited state molecular dynamics technique [Tretiak *et al.* (2002); Tretiak and Mukamel (2002)] recently developed and successfully applied to many conjugated molecular materials [Mukamel *et al.* (1997); Tretiak *et al.* (2000b); Franco and Tretiak (2004); Tretiak *et al.* (2003); Wu *et al.* (2006)]. This method makes simulations of exciton-vibrational dynamics in very large systems of up to one thousand atoms possible, while retaining the necessary quantitative accuracy. This is achieved by combining three techniquesi. (i) Reliable semiempirical all-valence approaches such as Austin Model 1 (AM1) [Dewar *et al.* (1985)] or Zerner's Intermediate Neglect of Differential Overlap (ZINDO) [Zerner (1996); Baker and Zerner (1991)] models. Semiempirical methods treat curvature and vibrational effects at significantly reduced computational complexity of the Hamiltonian, compared to *ab initio* approaches. (ii) A TDHF approximation used to address essential electronic correlations and excitonic effects. The TDHF combined with Krylov-subspace algorithms [Stratmann *et al.* (1998)] in the Collective Electronic Oscillator (CEO) code [Mukamel *et al.* (1997)], makes it possible to calculate hundreds of molecular excited states with only moderate numerical expense. (iii) Finally, analytic gradients of the excited-state potential energy surfaces [Furche and Ahlrichs (2002)] in the ESMD package allow investigation of vibrational phenomena, excited state optimizations, and adiabatic dynamics [Tretiak *et al.* (2002, 2003)].

4.2.1 *Hamiltonian Model and Electronic Correlations*

To study the exciton-vibrational phenomena in SWCNTs, we employed the following computational strategy. Ground state optimal geometries were obtained using the AM1 semiempirical Hamiltonian [Dewar *et al.* (1985)] at the HF level. The AM1 approach was specifically designed for this purpose and was widely applied to calculate ground [Stewart (2000)] and excited state [Tretiak *et al.* (2000a); Franco and Tretiak (2004)] properties of many molecular systems. This includes chemical energies, geometries, dipoles, excitation energies and polarizabilities. The semiempirical approximation restricts the basis set to valence orbitals of Slater-type. This limits the number of computed Hamiltonian matrix elements and allows storage of all of them in memory, instead of recalculating them when needed, as is commonly done in *ab initio* computations [Szabo and Ostlund (1989)]. This methodology makes semiempirical techniques significantly easier and faster yet allows for accurate description of a broad range of electronic phenomena. For example, no assumptions to the vibrational properties and curvature mediated σ and π interactions are necessary, as the AM1 Hamiltonian has these effects built into the dependence of its matrix elements on the nuclear positions. This constitutes an important advantage over the simpler tight-binding π-electron empirical approximations such as PPP [Zhao and Mazumdar (2004a); Zhao *et al.* (2006)] or SSH [Perebeinos *et al.* (2005c, 2004, 2005a)] models.

Optimized ground-state geometries provide input structures for excited state calculations performed using the CEO code combined with the AM1 Hamiltonian. The CEO approach, described in detail elsewhere [Tretiak and Mukamel (2002); Mukamel *et al.* (1997)], solves the equation of motion for the single-electron density matrix [Davidson (1976)]

$$\rho_{mn}(t) = \langle \Psi(t) | c_m^\dagger c_n | \Psi(t) \rangle \,, \tag{4.1}$$

of a molecule driven by an external electric field using the TDHF approximation for the many-electron problem [Thouless (1972); Ring and Schuck (1980)]. Here $|\Psi(t)\rangle$ is the many-electron wavefunction (time-dependent single Slater determinant driven by an external field), $c_m^\dagger$ (c_m) are creation (annihilation) operators, and the indices m and n refer to known basis functions (e.g., atomic orbitals, AOs, in the site representation). The TDHF equations of motion follow the evolution of the reduced density matrix representing the molecule driven by an external field $\rho = \rho_g + \delta\rho$ where the ρ_g is ground-state density matrix. The matrix elements of $\delta\rho$ represent the changes in these quantities induced by the external electric field. Within this theoretical framework, the changes induced in the density matrix by an external field are expressed as linear combinations of the electronic transition densities $\{\xi_\eta\}$[Tretiak and Mukamel (2002); Mukamel *et al.* (1997)]. These are defined as

$$(\xi_\eta)_{mn} = \langle\eta|c_m^\dagger c_n|g\rangle, \tag{4.2}$$

and reflect the changes in the electronic density induced by an optical transition from the ground state $|g\rangle$ to an excited state $|\eta\rangle$. Here $|g\rangle$ and $|\eta\rangle$ correspond to ρ_g and $\rho_g + \delta\rho$.The transition densities (or electronic modes) are, in turn, the eigenfunctions of the two-particle Liouville operator L from the linearized TDHF equation of motion [Thouless (1972); Ring and Schuck (1980); Linderberg *et al.* (1972); Tretiak and Mukamel (2002)]

$$L\xi_\eta = \Omega_\eta \xi_\eta, \tag{4.3}$$

where the eigenvalues Ω_η are electronic $|g\rangle \rightarrow |\eta\rangle$ transition energies. Eigenvalue problem Eq. 4.3 may be written in the matrix form as [Thouless (1972); Linderberg and Öhrn (1973); Tretiak and Mukamel (2002)]

$$\begin{pmatrix} A & B \\ -B & -A \end{pmatrix} \begin{bmatrix} X \\ Y \end{bmatrix} = \Omega \begin{bmatrix} X \\ Y \end{bmatrix}, \tag{4.4}$$

which is known as the first-order Random Phase Approximation (RPA) eigenvalue equation [Linderberg *et al.* (1972); Tretiak and Mukamel (2002)]. The analogous eigenvalue-problem is solved within an adiabatic TDDFT framework [Casida (1995); Onida *et al.* (2002); Stratmann *et al.* (1998)]. Here X and Y are particle-hole and hole-particle components of the transition density $\xi = \left[\begin{smallmatrix} X \\ Y \end{smallmatrix}\right]$ in the molecular orbital (MO) representation, respectively. The matrix A is Hermitian and identical to the Configuration Interaction (CI) singles matrix (CI singles is also known as Tamm-Dancoff approximation). The matrix B represents higher order electronic correlations included in the TDHF approximation and known as a de-excitation operator. The direct diagonalization of an operator L in Eq. 4.4 is a potential computational bottleneck of the excited-state calculations. The CEO procedure circumvents this problem using numerically efficient Krylov space algorithms (e.g., Lanczos or Davidson) [Davidson (1975); Stratmann *et al.* (1998); Tretiak and Mukamel (2002)]. This is possible since the action of the TDHF operator L on an arbitrary single electron matrix ξ can be calculated on the fly without constructing and storing the full matrix L in memory (so-called direct approach) [Rettrup (1982); Stratmann *et al.* (1998); Tretiak and Mukamel (2002)]. Subsequently, the computation of excited states is not substantially more numerically demanding than the ground state calculations. The TDHF approximation accounts for essential electronic correlations, such as electron-hole interactions, including some additional higher order terms [Thouless (1972); Ring and Schuck (1980); Pines and Bohm (1952)], which are sufficient for a reasonably accurate calculations of UV-visible spectra in many extended organic molecular systems [Tretiak and Mukamel (2002)]. It has the advantage of being size consistent, in contrast to many truncated CI techniques [Szabo and Ostlund (1989)].

4.2.2 *Exciton-Vibrational Dynamics and Relaxation*

The vibrational dynamics after initial photoexcitation is followed using the ESMD approach [Tretiak *et al.* (2002)], which calculates classical nuclear trajectories on the excited-state adiabatic potential energy hypersurface. This initial photoexcitation (hot exciton) is allowed to evolve along the molecular excited state potential energy surface $E_e(\mathbf{q})$, illustrated in Fig. 4.2, according to the Newtonian equations of motion for the nuclear degrees of freedom:

$$M_\alpha \frac{\partial^2 q_\alpha}{\partial t^2} + b\frac{\partial q_\alpha}{\partial t} = F_\alpha = -\frac{\partial E_e(\mathbf{q})}{\partial q_\alpha}, \qquad \alpha = 1, \cdots, 3N-6, \tag{4.5}$$

using a numerical velocity Verlet finite difference algorithm [Allen and Tildesley (1987)]. Here q_α and M_α represent the coordinates and the mass of one of the $3N-6$ vibrational normal modes (N being the total number of atoms in the molecule). The forces F_α are obtained as analytical derivatives of an exited-state energy $E_e(\mathbf{q})$ with respect to q_α [Furche and Ahlrichs (2002)] (see Fig. 4.2). We follow the dynamics of all $(3N-6)$ nuclear degrees of freedom of the molecule. The excited-state potential energy surface $E_e(\mathbf{q})$ and the forces that enter into Eq. 4.5 are quantum-mechanically calculated using the CEO method. Namely, for each nuclear configuration $\mathbf{q}$, $E_e(\mathbf{q}) = E_g(\mathbf{q}) + \Omega_\eta(\mathbf{q})$. The ground state energy $E_g(\mathbf{q})$

and the vertical $|g\rangle \to |\eta\rangle$ transition frequency $\Omega_\eta(\mathbf{q})$ are both calculated with the CEO technique.

The ESMD code makes it possible to follow picosecond excited-state dynamics of quite large ($\sim$ 1000 atoms) molecular systems taking into account all their $3N - 6$ vibrational degrees of freedom. The simulations allow us to follow the gas-phase dynamics with vanishing damping ($b = 0$). In particular, by these means we can model coherent phonon dynamics in SWCNTs [Gambetta *et al.* (2006b)]. Imposing an effective viscous medium ($b \neq 0$) leads us to the excited state optimal geometry, when the system is propagated sufficiently long for equilibration to occur. This allows us to study 'cold' exciton properties, such as self-trapping phenomena [Tretiak *et al.* (2007b)].

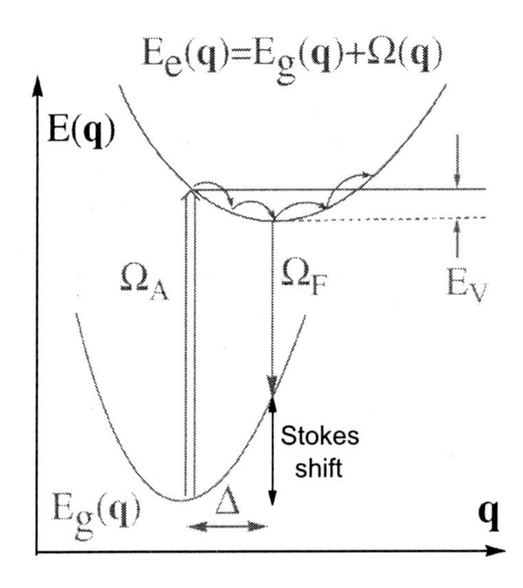

Fig. 4.2 Schematic representation of molecular dynamics propagation: The excited state energy $E_e(q)$ as a function of nuclear coordinates $\mathbf{q}$, displacements Δ, vibrational reorganization energy E_ν, vertical absorption Ω_A, fluorescence Ω_B, and Stokes shift $\Omega_A - \Omega_B$ frequencies.

The excited state vibrational relaxation is typically characterized by the dimensionless displacements Δ_α of each normal mode, as shown in Fig. 4.2, and respective Huang-Rhys factors $S_\alpha = \Delta_\alpha^2/2$. Geometry optimizations of a higher-lying excited state (e.g., the second exciton E_{22} of SWCNTs) becomes nearly impossible because of a significant density of states, level crossings and numerical expense involved. However, a rough approximation can be applied to estimate exciton-phonon effects. This assumes the same set of vibrational normal modes for both ground and excited states. Subsequently, the gradient of the electronic transition energy $E_e(\mathbf{q})$ along the vibrational coordinate of interest q_α provides the exciton-phonon coupling element $V_\alpha^{ep} = \frac{\partial E_e(\mathbf{q})}{\partial q_\alpha}$, and the approximate Huang-Rhys factor for an excited state of interest is given by

$$S_\alpha = \frac{(V_\alpha^{ep})^2}{\omega_\alpha^2}, \tag{4.6}$$

We previously found that such approximation works well in the case of SWCNTs due to their rigid structure. Calculated displacements using 'exact' (optimal geometries) and 'approximate' (obtained by Eq. (4.6)) approaches for E_{11} agree within about 10% for RBM for selected nanotubes, which translates to about 20% differences in the Huang-Rhys factors [Shreve *et al.* (2007a)].

4.2.3 *Real-Space Analysis*

During the photoexcited dynamics, the molecular geometry gets distorted and this, in turn, induces strong changes in the electronic wavefunction. To connect these structural changes with the distinct dynamics of the underlying photoinduced electron-hole pairs we use a two-dimensional real-space analysis [Tretiak and Mukamel (2002); Mukamel *et al.* (1997)] of the calculated transition densities ξ_η (Eq. 4.2). The diagonal elements of the transition densities $(\xi_\eta)_{nn}$ represent the net charge induced in the n-th AO by an external field. The off-diagonal elements $(\xi_\eta)_{mn}$ $(m \neq n)$ represent the joint probability amplitude of finding an electron and a hole located on the m-th and n-th AOs, respectively.

To obtain a two-dimensional real-space display of these modes, we coarse grain them over the various orbitals belonging to each atom. In practice, the hydrogens are omitted because they weakly participate in the delocalized electronic excitations. For other atoms, we use the following contraction: the total induced charge on each atom A is given by:

$$(\xi_\eta)_A = \left| \sum_{n_A} (\xi_\eta)_{n_A n_A} \right|, \tag{4.7}$$

whereas an average over all off-diagonal matrix elements represents the effective electronic coherence between atoms A and B,

$$(\xi_\eta)_{AB} = \sqrt{\sum_{n_A m_B} [(\xi_\eta)_{n_A m_B}]^2}. \tag{4.8}$$

Here the indices n_A and m_B run over all atomic orbitals localized on atoms A and B, respectively. The size of the resulting matrix $(\xi_\eta)_{AB}$ is now equal to $N' \times N'$, N' being the number of atoms in the molecule without hydrogens. Contour plots of $(\xi_\eta)_{AB}$ provide a real-space picture of electronic transitions by showing accompanying motions of optically induced charges and electronic coherence [Tretiak and Mukamel (2002); Mukamel *et al.* (1997)]. This is illustrated schematically by Fig. 4.3: the coordinate axes label atoms and indices A and B of matrix $(\xi_\nu)_{AB}$ run along the y and x axes, respectively. Two characteristic lengths are of relevance in these plots. The diagonal size of the non-zero matrix elements L_D reflects the degree of localization of the optical excitation (the position of the center of mass of the electron-hole pair). The largest off-diagonal extent of the nonzero matrix area (coherence length L_C) measures the maximal distance between the electron and hole (the exciton size). These L_D and L_C cross-sections are schematically shown in Fig. 4.3.

4.2.4 *Simulation Details*

The initial structures of fourteen SWCNTs of various length have been generated using TubeGen 3.3 [Frey and Doren (2005)]. Unsaturated chemical bonds at the open tube ends (see Fig. 4.4) have been capped with hydrogen atoms and methylene (CH_2) groups to remove mid-gap states caused by dangling bonds. We have checked several different configurations of H- and CH_2-capping for each tube. Among these configurations, the final capping is accepted onlyif it provides a smooth dependence of HF energy gap on a length of a SWCNT, approaching the limit of an infinite tube (simulated by imposing one dimensional periodic boundary conditions). Moreover, both finite size and periodic system approaches result in the nearly identical optimal geometries in the bulk of the tube (1-2 nm away from the tube ends). Subsequently, the tube ends do not introduce artifacts into tubes geometries and their electronic structure. All simulated molecular systems are oriented along z-direction, have a finite length (varied from 1 to 10 nm) and comprise several repeat units (see Table 4.1). The maximum length of nanotubes is chosen to be significantly larger than the diameter of the tube and characteristic exciton sizes (about 5 nm, see below). If these conditions are satisfied, the finite-size 1D systems are expected to reproduce the properties of the infinite-size systems.

Furthermore, we used the MOPAC-2002 code [Stewart (2000)] and AM1 model to obtain ground state optimal geometries, heats of formation (chem-

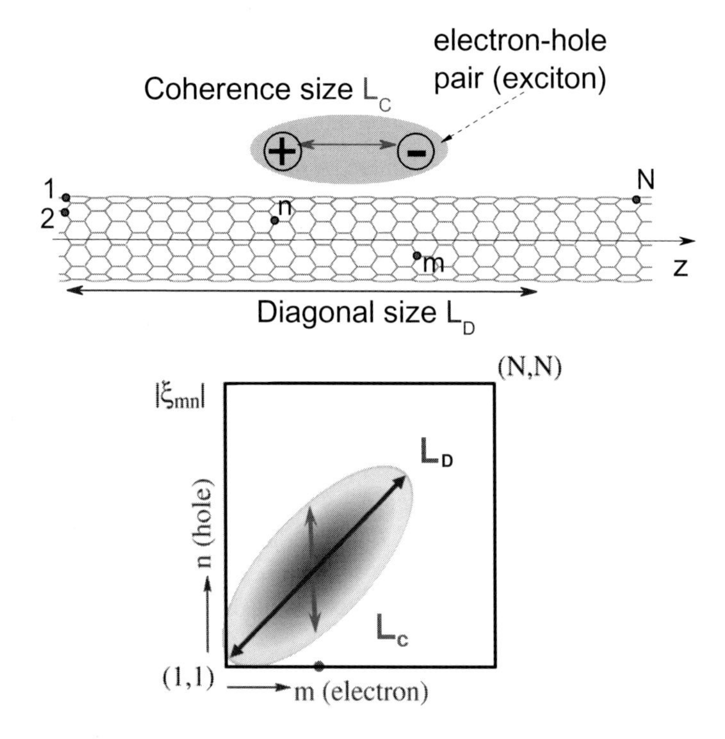

Fig. 4.3 Two-dimensional representation and physical significance of electronic modes. Each mode ξ_ν is an $N \times N$ matrix, N being the number of atoms. The contour plot provides a direct real-space connection between the optical response and motions of charges in the molecule upon optical excitation. The x-axis represents an electron on a atom n, and the y-axis describes a hole on atom m. The numbering of atoms of SWCNT is in the increasing oder of atomic coordinate z, chosen parallel to the tube axis. The incident light moves an electron from some occupied to an unoccupied orbitals, creating an electron-hole pair (or an exciton). The state of this pair can be characterized by two lengthscales. First, the distance between electron and hole (i.e., how far the electron can be separated apart from the hole. This coherence size L_C is the size of the density matrix along the antidiagonal direction. The second length L_D describes the exciton center of mass position (i.e., where the optical excitation resides within the molecule), presented as the diagonal direction along ξ_ν.

ical energies), and vibrational normal modes of all finite size tubes (see Fig. 4.4). The vertical transition frequencies from the ground state to the singlet excited states, their oscillator strengths, and transition density matrices were then computed with the CEO procedure.

The SWCNT singlet excitonic states E_{ii} formally correspond to van Hove singularities [Saito *et al.* (1998a)]. However, the ratio between ex-

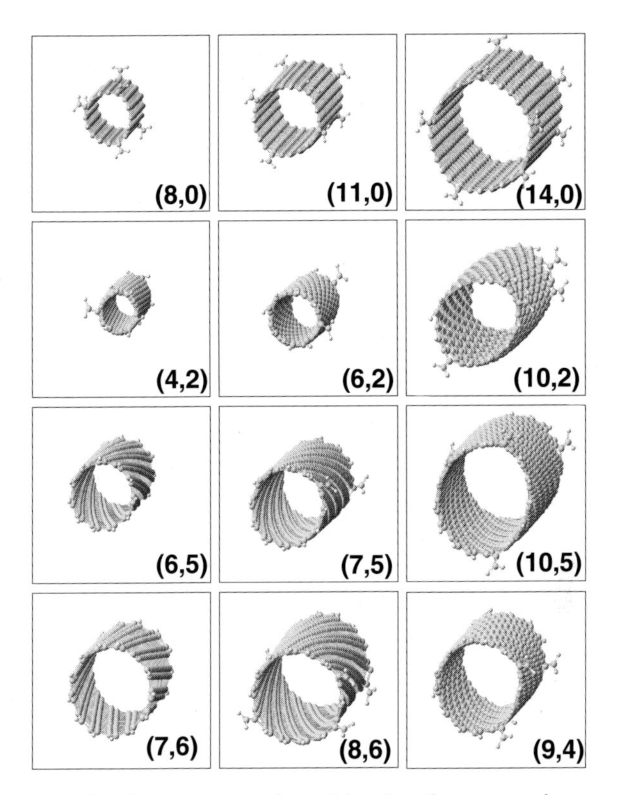

Fig. 4.4 Optimal molecular structure of considered carbon nanotubes, presented in perspective depth view for visual comparison of tubes' diameters and chiralities. The upper row displays zigzag tubes, lower rows present tubes with increasing chiralities. All tubes are finite and chosen to have approximately the same length around 7-8 nm in Z-direction. Electronic dangling bonds are terminated with either -H or -CH2 capping agents.

citonic energies violates predictions of conventional one electron theories due to electronic correlation effects [Kane and Mele (2004)]. Excited states of the finite-size molecules are discreet in contrast to the band structure obtained from infinite tube calculations [Spataru *et al.* (2004); Perebeinos *et al.* (2004); Chang *et al.* (2004)]. Each excitonic band is a manifold of closely spaced levels. The lowest state typically corresponds to the tightly bound exciton E_{ii}, which collects nearly all of the oscillator strength from its parent band. Conjugated polymers provide well-explored example of typical 1-D electronic structures that share many striking similarities with SWCNTs case [Zhao and Mazumdar (2004a); Zhao *et al.* (2006); Tretiak

Tube	Diameter (nm)	Unit length (nm)	max number of units	max. length (nm)	Number of atoms
(4,2)	0.414	1.13	9	9.80	568
(6,2)	0.565	1.54	7	10.10	744
(8,0)	0.626	0.43	25	10.54	812
(6,5)	0.747	4.07	2	8.14	748
(7,5)	0.817	4.45	2	8.98	898
(11,0)	0.861	0.43	21	8.96	940
(10,2)	0.872	2.38	2	4.63	524
(7,6)	0.886	4.81	2	9.62	1040
(9,4)	0.903	4.92	1	4.86	562
(8,6)	0.952	2.59	3	7.82	920
(10,5)	1.039	1.13	7	7.79	1008
(14,0)	1.100	0.43	17	7.17	972
(16,0)	1.270	0.43	15	6.32	980
(19,0)	1.505	0.43	12	5.05	936

(2007a); Tretiak *et al.* (2007b)].

The computed singlet states are strongly delocalized $\pi - \pi$ excitations, which are optically accessible. Apart from the lowest E_{11} excitation, E_{22} excitation usually lies within the first 100 electronic states in all of the finite-size structures we considered. E_{11} and E_{22} calculations can be done routinely with the CEO procedure. In contrast, calculations of about 500-700 states are required to access E_{33} excitonic band because of the high density of states in the high-frequency spectral region. This becomes numerically expensive and memory-demanding, since the CEO computational effort typically scales linearly with the number of excited state requested. Among hundreds states that separate E_{11} and E_{22} transitions, there are several optically forbidden (completely dark) exciton states with zero transition dipole moments and a few semi-dark excitons with relatively small transition dipole moments, which are nearly isoenergetic to the allowed E_{11} state in SWCNTs.

The relative ordering of bright and dark states manifest the poor fluorescence efficiency of the CWCNTs [Zhao and Mazumdar (2004a)]. However, we found that state ordering is highly method-dependent in semiempirical approaches. Thus, CEO calculations based on the AM1 Hamiltonian provide the bright E_{11} exciton as the lowest state in the first band of SWCNTs independently on a tube chirality and diameter. In contrast, CEO coupled with ZINDO Hamiltonian predicts the bright E_{11} exciton being the second low-energy state in the band of tubes with diameter larger than 0.7 nm, while for narrower tubes, there are 7-10 dark states laying lower than the bright one [Kilina *et al.* (2008)]. TDDFT methodology was recently applied to address the state ordering, their delicate energetics, and optical activity [Scholes *et al.* (2007)]. The results agree well with experimental data and kinetic modeling [Scholes *et al.* (2007)] showing low-energetic dark excitons placed below the optically active one.

To sort discrete states into excitonic bands, we use real-space analysis to examine the nodal structure along diagonals of the matrices. Delocalized excited states can be considered as standing waves in quasi-one-dimensional structures, and the nodes are related to the quasi-particle (exciton) momenta [Tretiak *et al.* (2000b)]. Consequently, the zero-node state is associated with the $k = 0$ momentum exciton in the infinite chain limit, which may be optically allowed. For linear molecules, this correspondence was explored in the context of electron-energy-loss spectroscopy [Chernyak *et al.* (2001b,a)]. Also, such an approximation has been recently applied to the conjugated polymers and dendrimers [Wu *et al.* (2006)], allowing accurate modelling of excited-state dynamics in arbitrary branched molecular structures. It is important to note that nanotubes can not be regarded as pure 1D systems, due to their circumferential dimension: The curvature and chirality of tubes give rise to some interference between excitonic bands. Here, we focus only on zero-node states in SWNTs. States having one or more nodes are not considered, because they are assumed to belong to the same band as the respective zero-node exciton and originate from the finite tube sizes considered. Below the bright E_{22} exciton, there are typically 9-11 states, including the bright E_{11} state, whose transition densities do not have nodes.

To understand excited state vibrational relaxation we further optimize the geometry of the first optically allowed E_{11} excited state in the space of all vibrational coordinates $\mathbf{q}$ using the ESMD approach, which is related to the nanotube emission properties.

4.3 Results and Discussion

4.3.1 *Heat of Formation and Energy Gaps*

Figure 4.4 shows the structures corresponding to the minimum energy at the ground state for most types of SWCNTs studied and allows one to visualize tube sizes and chiralities. The geometric parameters of all calculated SWCNT species (ten chiral and five zigzag tubes) are given in Table 4.1. The heat of formation per carbon atom for the fourteen SWCNTs at their maximum length (calculated with the MOPAC-2002 code [Stewart (2000)]) is presented in Fig. 4.5. The heat of formation is a standard enthalpy of formation or the enthalpy change to form a mole of molecular compound at 25^oC from its elements. Figure 4.5 shows that tubes with smaller diameters ($\sim$ 0.5 nm), and, consequently, larger curvatures, have much larger heats of formation than wider tubes. The aromatic rings in nanotubes are no longer planar but distorted due to curvature. This leads to deviation from energetically preferable ideal sp^2 hybridization and π-delocalization of the graphene sheet. The curve in Fig. 4.5 becomes more flat for tubes with diameters of about 1 nm, which is a common size of SWCNTs used in experiments. Such a strong difference in the heat of formation between small and large SWCNTs explains the fact that very narrow tubes are much harder to synthesize and are rarely observed in experiments [Li *et al.* (2001)]. Tube-geometry deformation related to the curvature leads to a number of electronic phenomena, such as Peierls distortion and gap opening in metallic tubes [Saito *et al.* (1998a); Figge *et al.* (2001, 2002); Bohnen *et al.* (2004a); Dubay *et al.* (2002)]. Below we discuss these geometric effects on the excitonic structure of semiconducting SWCNTs [Tretiak *et al.* (2007b); Gambetta *et al.* (2006b)].

The one-electron uncorrelated AM1/HF energy gap (the energy difference between HOMO and LUMO) is shown in Fig. 4.6(a) and (c). The respective calculated correlated CEO gaps are displayed in Fig. 4.6(b) and (d). Figure 4.6(a) demonstrates the effects of the finite length of the SWCNTs on their electronic structures. The smooth dependence of the energy gap on the reciprocals of the tube length, roughly following $1/L$ scaling law, assumes that the terminated tube ends have saturated dangling bonds and do not introduce their own electronic states to the tube gap even in very short tubes. This can also be directly confirmed by plotting the relevant HOMOs and LUMOs (not shown). Subsequently, the energy gap of the capped finite SWCNTs asymptotically approach the infinite tube limit

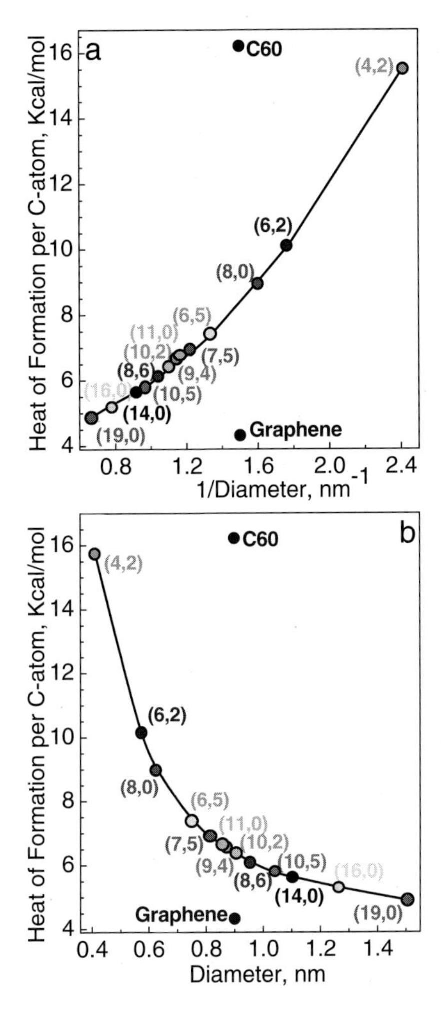

Fig. 4.5 Variation of calculated heat of formation per carbon atom as a function of **(a)** inverse tube diameter and **(b)** tube diameter, D. Wider tubes have the lower heat of formation and are more frequently observed in experiments. For comparison, calculated heats of formation for fullerene C_{60} (D=0.71 nm) and graphene are shown as well.

(Fig. 4.6(a)).

The dependence of the calculated HF energy gaps on a tube diameter and its comparison with experimental data are shown in Fig. 4.6(c). For uniform comparison, all calculated values are shown for tubes with roughly the same length (6.5 – 7 nm). The calculated dependence demonstrates

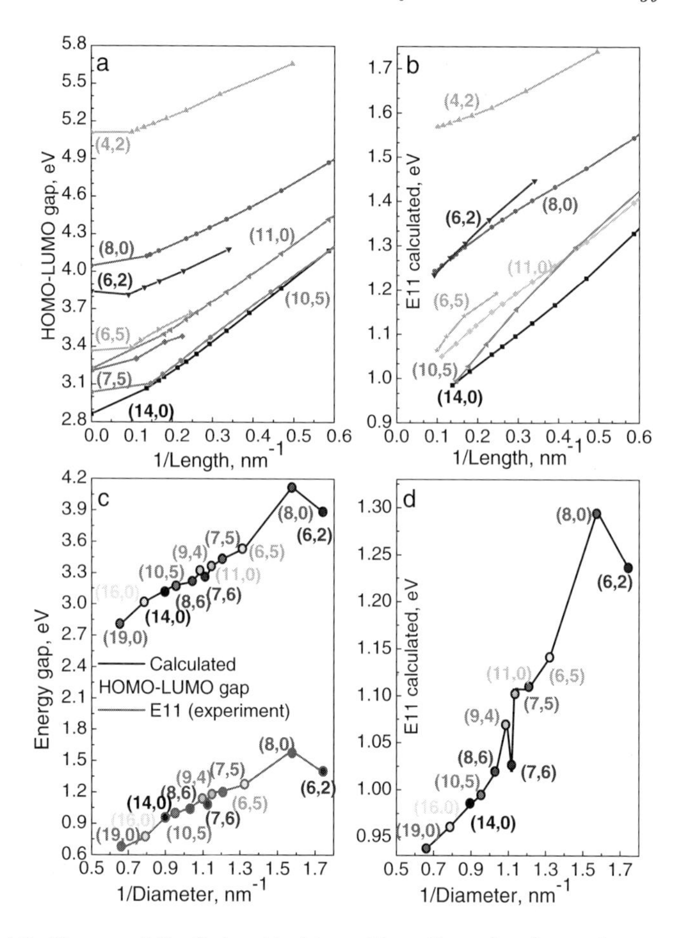

Fig. 4.6 Energy of the first optical transition, E_{11}, of various tubes as a function of inverse tube length ((**a**)) and ((**b**)), and inverse tube diameter ((**c**)) and ((**d**)). Panels (**a**) and (**c**) show scaling of uncorrelated HF/AM1 gap (HOMO-LUMO). Panels (**b**) and (**d**) show scaling of E_{11} transition energy obtained with CEO (TDHF/AM1) method, which incorporates essential electron-hole interaction (excitonic effects). Experimental data for the first optical transition from the literature [Bachilo *et al.* (2002); Weisman and Bachilo (2003)] are given by the red line in (c), for comparison. Overall, the calculated energy gaps reproduce trends seen in measured energies. HF/AM1 level overestimates experimental values by $\sim$ 2eV, which attributes to the lack of electronic correlations. The CEO results coincide well with experimental data with an accuracy of about 10-15%.

a surprisingly good agreement with experiment. Mod-1 and mod-2 tube families defined as $\text{mod}(n - m, 3) = 1$ or 2, respectively, are clearly distinguished by red-shifted energies of mod-2 tubes (7,6) and (6,2). Thus, the qualitative behavior of both calculated and experimental energies versus tube diameter strongly coincide. However, calculated HF energy gaps are uniformly shifted up by about 2 eV compared to the experimental values for all considered SWCNTs. The over-estimation of the energy gap is a common feature of the HF calculations and can be corrected by the inclusion of electronic correlations (Coulomb interaction between excited electrons and holes). Calculated with the CEO, correlated E_{11} optical transition energies at TDHF/AM1 level are red-shifted by $\sim$ 2 eV compared to the respective HF values; however, all scaling trends are preserved, as illustrated in Fig. 4.6(b) and (d). Compared to the corresponding experimental data, the CEO results exhibit a red-shift up to 0.15 eV for the tubes with a diameter less than 1 nm. This red shift originates from the AM1 Hamiltonian, which also often underestimates energy gaps for polymers [Tretiak *et al.* (2000a)]. In contrast, the tubes with diameters larger than 1 nm show small blue-shifts of transition energies from experimental values. This deviation grows with tube diameter because the characteristic excitonic sizes enlarge with diameter increase (see discussions below). Subsequently, the tube end effects start to introduce quantum confinement along the tube in the widest tubes (16,0), and (19,0), where computations are limited to relatively short segments ($\sim$ 7 nm in length).

Finally, we discuss emerging size-scaling laws for calculated energy gaps. There has been substantial effort in carbon nanotube research to develop scaling relationships between tube diameter and excitation energy (commonly used for Kataura plots) well beyond the simple $1/D$ dependence. Modern empirical scaling laws accurately reproduce experimental data for a broad range of tube diameters and account for chirality effects [Weisman and Bachilo (2003); Fantini *et al.* (2004b); Araujo *et al.* (2007)]. Similar work was done in the conjugated-polymer field to extrapolate scaling of bandgap energies of finite oligomers to the polymer limit [Hutchison *et al.* (2003); Gierschner *et al.* (2007)]. However, an inverse approximate relationship $1/L$ between an energy and a length, which may be rationalized with a free-electron model [Kuhn (1948b)], does not recover a finite band gap at infinite length. Indeed in Fig. 4.6(a), we observe that calculated energy gaps of the infinite-length SWCNTs demonstrate saturation effects as the gap becomes roughly constant at some large length. Instead of $1/L$ dependence, more sophisticated relationships based on the interacting oscillator models were introduced by Kuhn and applied to organic dyes [Kuhn (1948b,a, 1949)]. This approach was later extended within an exciton formalism [Simpson (1955); Chang *et al.* (2000)]. In particular, the following scaling relationship has been successfully used for conjugated polymer fitting in a recent review [Gierschner *et al.* (2007)]:

$$\Delta E = A\sqrt{1 + B\cos\left(\frac{\pi}{N+1}\right)}. \tag{4.9}$$

Here N is the number of repeat units in the segment and A and B are fitting constants. We use Eq. (4.9) to fit both calculated uncorrelated and correlated gaps of narrow (4,2) and relatively wide (14,0) tubes. The results, presented in Fig. 4.7, show a good fit for E_{11} gaps in short and long segments, as well as the infinite tube limits, and clearly demonstrate deviation from $1/N$ relationship at long tube lengths.

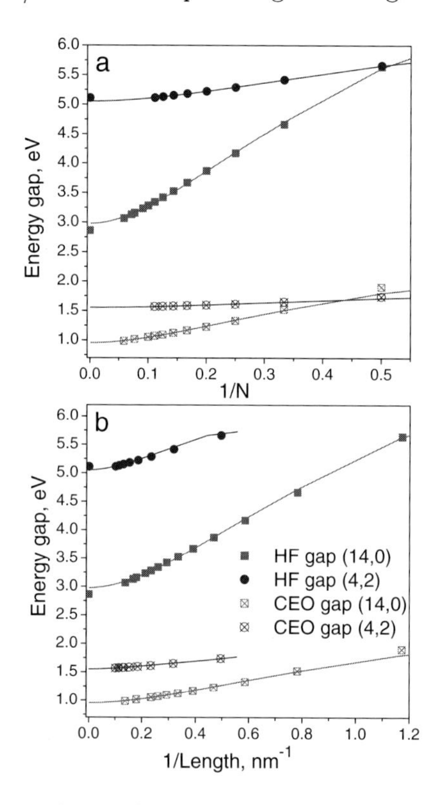

Fig. 4.7 Kuhn fit of the calculated uncorrelated and correlated energy gaps of (14,0) and (4,2) finite tubes using Eq. (4.9) as a function of **(a)** an inverse number of repeat units and **(b)** an inverse tube length. Deviation from 1/N (1/L) relationship at long tube lengths is clearly seen.

4.3.2 *Cross- and Parallel-Polarized Excitons from the Lowest Band*

To study E_{11} and E_{22} exciton structures, we calculate up to the 100 lowest singlet excited states for each SWCNT from Table 4.1 at the ground state (GS) optimal geometry. E_{11} and E_{22} states have characteristically strong transition dipole moment directed along the tube axis (z-polarized), which results in distinct peaks in the SWCNT's linear absorption spectra probed by a variety of spectroscopic techniques [Bachilo *et al.* (2002); Ma *et al.* (2005b); Korovyanko *et al.* (2004a); Manzoni *et al.* (2005b); Wang *et al.* (2005a); Maultzsch *et al.* (2005); Zhao *et al.* (2006)]. Between the E_{11} and E_{22} transitions, there are many optically forbidden dark states and a few weakly allowed excitons, all representing delocalized transitions. Each excitonic band is a manifold of closely spaced levels.

As an example, Fig. 4.8 presents the contour plots of the transition density matrix corresponding to the zero-node states from the first excitonic manifold of the (10,5) SWNT. Other SWNTs we have studied exhibit similar trends and are not shown. Each transition density matrix is labeled by a number according to the energy of the photoexcited state with respect to the ground state. For comparison, the excitation energies and the dominant x-, y-, or z-component of the oscillator strength (f) are also shown on each plot. These plots can be considered as topographic maps that reflect how the single-electron reduced density matrix changes upon molecular photoexcitation. In all color panels the axes correspond to carbon atoms, whose coordinates are labeled along the tube axis. Consequently, these plots mostly show the distribution of the excitonic wave function along the tube. Thus, the nine displayed first excitons of the (10,5) tube are all delocalized transitions with respect to the tube length, and, mainly, have weak or zero oscillator strength, except the second (bright) E_{11} state.

The respective slices along the matrix diagonal (L_D direction) for all nine excitons are presented in Fig. 4.9. These characterize the distribution of an excitonic wave function along the length of the tube. The presence of only one maximum in the distribution indicates the zero-node character of the excitonic wavefunction. Both matrix contour plots and diagonal L_D slices (Fig. 4.8 and Fig. 4.9) demonstrate that the center of mass of all nine first-band excitons is spread over the entire (10,5) tube, which is typical for all studied SWNTs. The amplitudes vanish at the tube edges, which reflects excitonic scattering (reflection) at the ends [Wu *et al.* (2006)]. Such exciton delocalization patterns are identical to those observed in other quasi-one-

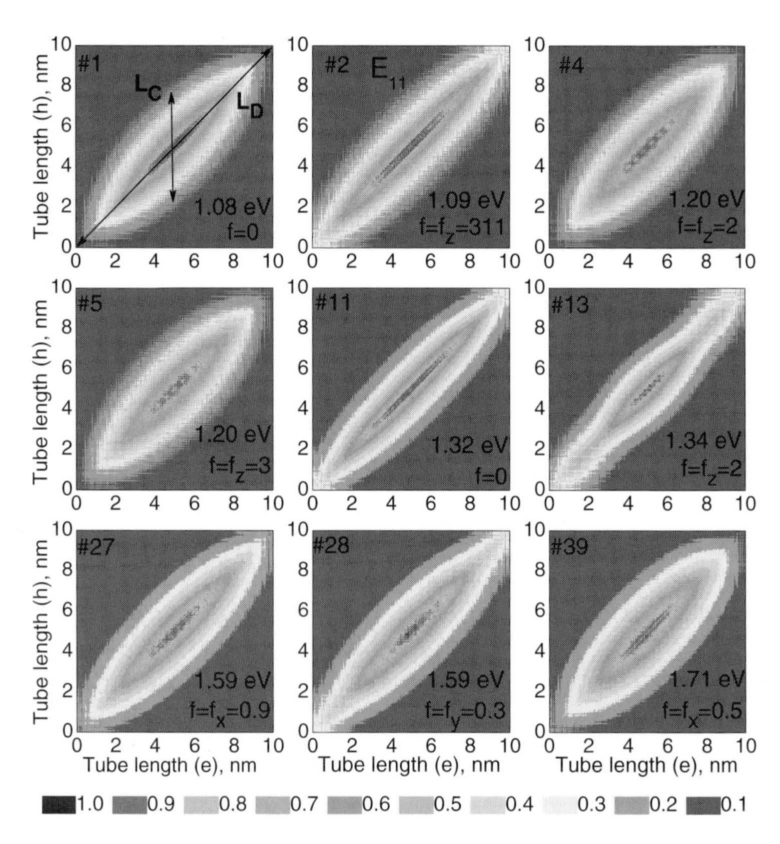

Fig. 4.8 Excitonic transition density matrices of the (10,5) tube plotted in Cartesian coordinates. The first nine zero-node excitonic states are presented by two-dimensional contour plots as a function of the electron (vertical-axis, nm) and hole (horizontal-axis, nm) coordinates along a tube axis. The color code is presented at the bottom. Excitons are labeled according to their order number with respect to the ground state. The transition energy and the dominant oscillator strength component (f) are shown for each excitonic state. The strongly allowed optically active (bright) exciton is marked by E_{11}. These plots reflect the distribution of the excitonic wave functions along the tube axis. For color reference, turn to page 154.

dimensional materials [Tretiak *et al.* (2007b); Tretiak and Mukamel (2002); Tretiak *et al.* (2002); Kilina and Tretiak (2007)]. Another important characteristic of the electronic excitation is the exciton coherence size L_C (maximal distance between electron and hole along the tube axis). The L_C cross-sections of the transition densities are shown in Fig. 4.8 and represent the probability distribution of an electron coordinate, when the position of a hole is fixed in the middle of a tube. Such representation typically reflects the delocalization and binding strength between electron and hole in SWNTs [Spataru *et al.* (2004); Araujo *et al.* (2007); Kilina and Tretiak (2007)]. Plots in Figs. 4.8 and 4.9 show that L_C is finite for all nine excitons. These correspond to tightly bound singlet excitons, with an excitonic size of 3.5-5 nm, depending on the state.

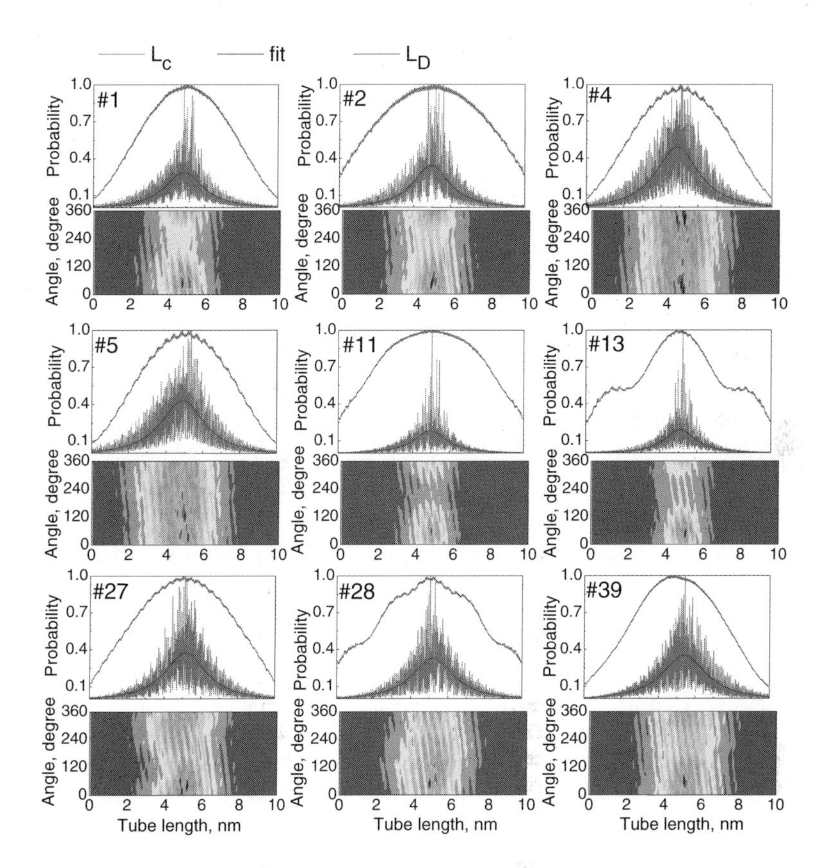

Fig. 4.9 Delocalization properties of nine zero-node excitonic states of the (10,5) tube, characterizing excitonic wave function distribution along the tube axis (upper sub-panels) and the tube circumference (lower sub-panels). The diagonal slice (L_D) of the transition density matrix represents the electron-hole pair center-of-mass position along the tube. Off-diagonal slices (L_C) indicate the electron coordinate along the tube, when the hole is fixed in the middle; thus, determining the maximal distance between an electron and a hole along the tube. This distribution is fitted by a Lorentzian function. The contour plots (lower sub-panels) represent the probability of finding the electron on the tube circumference when the hole is fixed in the middle of a tube. This probability is a function of position of the electron along the tube axis (X-axis, nm) and the electron position along the tube circumference (Y-axiz, degrees), measured as an angle $0 < \phi_i < 2\pi$. The color scheme is given in Fig. 4.8. For color reference, turn to page 155.

According to recent studies [Zhao and Mazumdar (2004a); Perebeinos *et al.* (2005c)], four parallel-polarized excitons are expected to form from the E_{11} transition of SWNTs due to near doubly-degenerate HOMO-LUMO transitions, three of which are dark. Our subsequent analysis will identify them as states 1, 2, 11 and 13 for the (10,5) tube in Figs. 4.8 and 4.9. By analogy, due to molecular orbital degeneracy, we expect to find eight distinct cross-polarized excitonic bands related to E_{12} and E_{21} transitions (see Fig. 4.1), most of which are also dark. Coulomb correlation effects are expected to lift the degeneracy from these states. Indeed, we are able to identify up to 4-6 such states depending on the nanotube considered. For example, Figs. 4.8 and 4.9 show 5 such states (4, 5, 27, 28, and 39). Higher energy cross-polarized states are hard to distinguish because of enhanced density of states and overlap with E_{22} excitonic bands. All these states have either a vanishing or a very weak oscillator strength. Note that, due to state mixing and open boundary conditions, some cross-polarized excitons might have a small amount of z-polarized transition dipole moment. Such state interference complicates the analysis of excitonic structure and the separation of z-polarized excitons from cross-polarized ones (e.g., states 4 and 5 in Figs. 4.8 and 4.9 have a small z-polarized component, whereas they represent cross-polarized excitations, as will be clarified later). Our calculations show that each considered tube has at least two states with nearly degenerate energies and dominating oscillator strength components directed along x and y (for example, states 27 and 28 in Fig. 4.8 and Fig. 4.9). Those states are weakly allowed cross-polarized excitations, for which degeneracy is attributed to cylindrical symmetry of the SWNTs. We emphasize that all parallel-polarized (E_{11}) and cross-polarized (E_{12} and E_{21}) states have very similar distributions of the excitonic wavefunctions along the tube axis and separations between the electron and hole (see Fig. 4.8 and Fig. 4.9), however, their angular momenta may differ substantially [Uryu and Ando (2006)]".

In order to characterize cross-polarized excitations, the distribution of the excitonic wavefunction along the tube circumference needs to be analyzed. In principle, transition density matrices contain both axial and azimuthal distributions of the excitonic wave function. However, these properties can not be visualized from the 2D color maps in Fig. 4.8 and from the L_D and L_C cross sections (Fig. 4.9 (upper panels)), which reflect rather an averaged distributions. The information regarding wave function delocalization with respect to the tube circumference can be extracted from the transition density matrices by fixing the position of one of the charges (either electron or hole) on a tube, while presenting the transition density matrix as a 2D function of electron position along the tube length and along the radial angular coordinate ϕ_i. The resulting contour plots of transition density matrix for the first nine zero-node excitons in the (10,5) tube are displayed in Fig. 4.9 (bottom panels). All nine excitons demonstrate a delocalized character along the tube circumference ϕ, in agreement with other theoretical findings [Maultzsch *et al.* (2005)]. However, states marked as 1, 2, 11, and 13 show a narrower distribution of the electron's density with respect to the radial angle, which, in addition to their smaller excitonic sizes, distinguishes these four excitons from the other states. These four excitons belong to the parallel-polarized parent E_{11} transitions, as evidenced by their oscillator strengths (e.g., the z-polarized bright exciton is among these four states).

A more accurate determination of cross- and parallel-polarized states can be obtained from the characteristics of electron-hole positions depending on the azimuthal angle ϕ_i using the natural *helical coordinate system.* The lowest unoccupied molecular orbitals (LUMOs) plotted for chiral (6,5) and zigzag (8,0) SWNTs in Fig. 4.10 (a) emphasize a well-defined relationship between the orbital density distribution along a tube surface and the natural symmetry of a chiral tube. As can be seen in Fig. 4.10 (a), the orbitals follow the chiral angle Θ between the tube axis and the chirality vector $\vec{C}_h = n\vec{a}_1 + m\vec{a}_2$, which determines the direction of the folding of a graphene layer to construct an (n, m) tube [Saito *et al.* (1998a)] (for illustration, see Fig. 4.10 (b)). Here $\vec{a}_{1,2}$ are lattice vectors of a graphene sheet. Thus, the chiral angle can be determined as $\cos\Theta = \frac{\vec{C}_h \cdot \vec{a}_1}{|\vec{C}_h| \cdot |\vec{a}_1|}$. An i-th carbon atom on a surface of the tube can be defined by its cylindrical coordinates: position along the tube axis z_i, and an azimuthal angle ϕ_i, as shown in Fig. 4.10 (b). To this end, we define a subset of carbon atoms with coordinates z_i and ϕ_i that satisfies the conditions

$$\Phi = \frac{z_i}{R\tan\Theta}, \tag{4.10}$$

$$|\Phi - \phi_i| \leq \delta\Phi, \tag{4.11}$$

where an arbitrarily small width parameter $\delta\Phi$ is roughly chosen as the angular distance between the two nearest carbons. Thus, for a given tube index (n, m) we can investigate the excitonic delocalization among atoms selected by Eqs. 4.10 and 4.11. For example, Fig. 4.11 (b) shows these atoms colored red for (10,5) tube. Cyan carbon atoms represent the replica of the 'red' line shifted by 180°. The cyan and red lines wrap around a tube

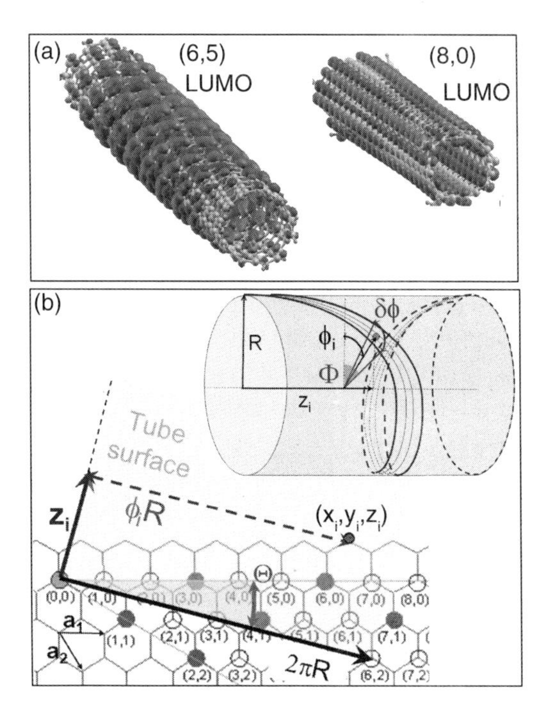

Fig. 4.10 (a) The lowest unoccupied molecular orbitals (LUMOs) of chiral (6,5) and zigzag (8,0) CNTs. The positive and the negative lobes of the wavefunction are shown by the red and blue colors, respectively. The orbital density distribution strongly follows the tube helicity. (b) Schematic presentation of the nanotube geometrical parameters. Vectors $\mathbf{a_1}$ and $\mathbf{a_2}$ are the unit vectors of the hexagonal graphene lattice. The chirality vector $\mathbf{C_h} = n\mathbf{a_1} + m\mathbf{a_2}$ (black arrow), chiral angle Θ, and the translational vector Z_i directed along the tube length (blue arrow) of a (6,2) tube are shown as an example. The inset shows natural helical coordinates of a chiral tube. All points on the surface area between the two blue lines satisfy Eq. 4.10 and 4.11. For color reference, turn to page 155.

from opposite sides. Comparison of transition density distributions between the 'red' and 'cyan' directions allows one to visualize how an exciton is mutually delocalized over both cross-laying tube sides.

Fig. 4.11 (a) shows the distribution of transition density matrix values along the 'red' line for nine zero-node states for (10,5) tube. To eliminate the artifacts of tube ends, we focus only on one central unit-cell of a tube. These plots demonstrate a nearly uniform distribution along the diagonal direction, which is expected from results presented in Fig. 4.9 (bottom panels). A similar picture is valid for the 'cyan' direction (not

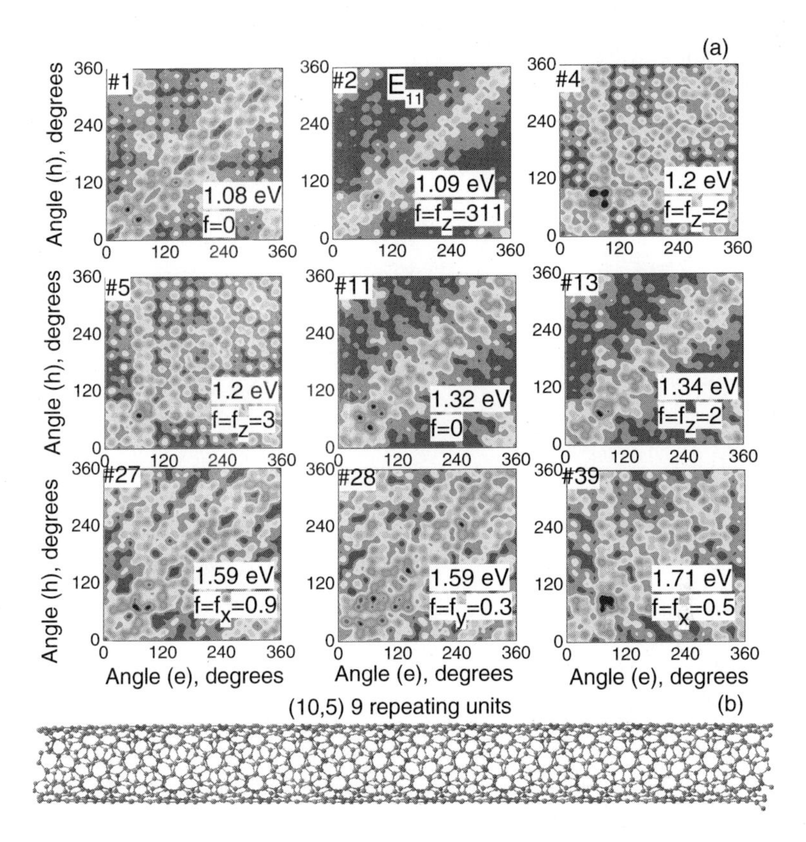

Fig. 4.11 (a) Contour plots of the transition density matrices corresponding to nine excitons of the (10,5) tube shown in Figs. 4.8 and 4.9 in the helical coordinate system schematically shown at bottom and described by Eqs. 4.10 and 4.11. Positions of the electron and hole are labeled with an azimuthal angle $0 < \phi_i < 2\pi$. The helical lines of the (10,5) tube are highlighted by red and cyan (orthogonal to the red line) colors on the tube surface. Four excitons numbered 1, 2 (bright state), 11, and 13, show much stronger localization with respect to the tube circumference, compared to other states. These four excitons are attributed to parallel-polarized E_{11} transitions, while the others are cross-polarized E_{12} and E_{21} excitons. The color scheme is given in Fig. 4.8. For color reference, turn to page 156.

shown) reflecting the delocalized character of the excitonic wave function with respect to the tube circumference. By analogy with L_C, the coherence azimuthal size of an exciton can be defined as the largest off-diagonal extent (L_{CA}) of the non-zero matrix area, presented in Fig. 4.11. It is clearly seen that L_{CA} is finite only for the four z-polarized excitons, while it is spread over the entire matrix for other states. The finite L_{CA} size indicates that z-polarized excitons originating from the E_{11} transition are more localized along the tube circumference as compared to the cross-polarized excitations arising from E_{12} and E_{21} transitions. Remarkably, transition density matrices analyzed in the helical coordinate systems have very similar patterns for (10,5) and other studied SWNTs (for example, see Fig. for (7,6) tube). This representation clearly distinguishes z-polarized states

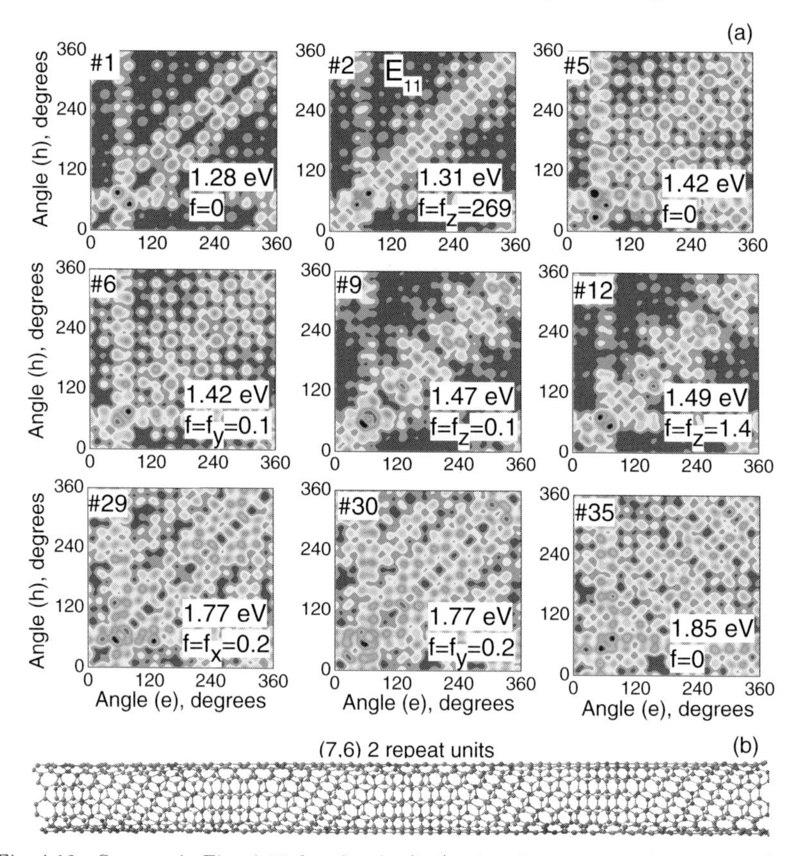

Fig. 4.12 Same as in Fig. 4.11, but for the (7,6) tube. Four excitons labeled 1, 2 (bright state), 9, and 12 are attributed to parallel-polarized transitions due to their localized character with respect to the tube circumference. The other states are cross-polarized excitons. For color reference, turn to page 157.

from cross-polarized ones. Among the four z-polarized excitons, one is optically active, two are completely dark, and one is weakly allowed with a dominant z-component in its oscillator strength. The other excitons are cross-polarized E_{12} and E_{21} states. The oscillator strength of these cross-polarized excitons is mostly suppressed. However, there are at least two semi-bright, nearly isoenergetic states with non-zero x- or y-components in their oscillator strength, which can be probed experimentally. Fig. 4.13 (a) represents numerically calculated polarized absorption spectra of one of the CNTs. The 1-node semi-dark cross-polarized excitons (such as 27 or 28 in Fig. 4.11) contribute to the main peak of x- and y-polarized absorption and marked by a red arrow in Fig. 4.13 (a). We checked that change in tube length affects insignificantly the position of this peak, while E_{11}, as well as small x- and y-polarized peaks, accompanying the main peak and originated from mixture of parallel and cross-polarized states with different momenta, are noticeably shifted.

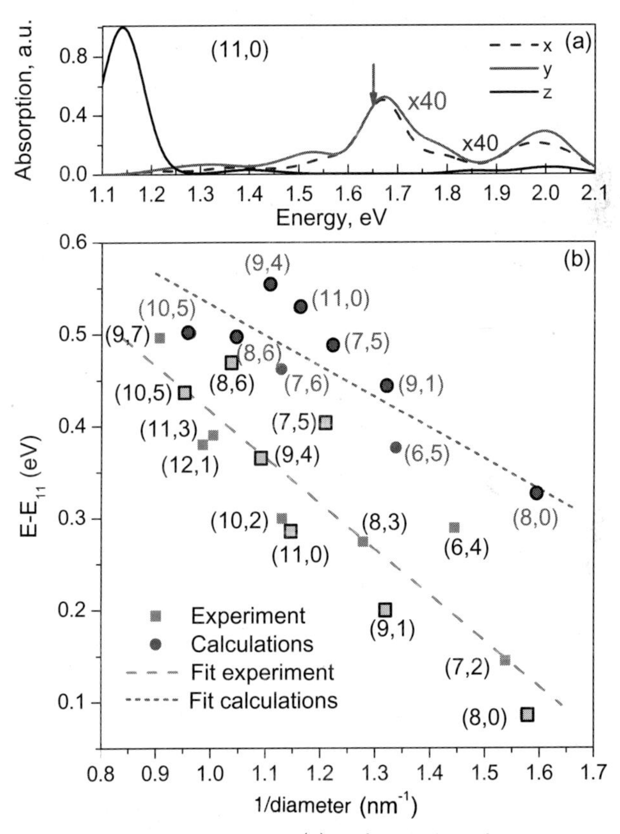

Fig. 4.13 Polarized Absorption spectra (a) and variation of excitation energy of semi-bright cross-polarized excitons with respect to the excitation energy of the first bright exciton with parallel polarization, as a function of the inverse tube diameter (b). The grey and red dashed lines show the fit of experimental and theoretical data using Eq. 4.12.

Fig. 4.13 (b) compares the experimentally measured (via Raman spectroscopy) [Luo *et al.* (2007)] and calculated energy splittings between the first bright E_{11} and a semi-bright cross-polarized E_{12} (E_{21}) excitations as a function of the tube diameter. Both experimental and calculated results demonstrate an increase in the splitting as tube diameter increases. This trend can be approximately linearly fit as

$$\Delta E = \Delta E_0 - S/d, \tag{4.12}$$

where S is the slope and d is the tube diameter. Numerical fitting gives $\Delta E_0^{exp} = 0.91$ eV, $\Delta E_0^{th} = 0.87$ eV, $S^{exp} = 0.50$ eV·nm, $S^{th} = 0.33$ eV·nm (in atomic units $S^{exp} = 0.35$, $S^{th} = 0.23$). Such scaling can be interpreted as a consequence of observed different excitonic delocalizations for the parallel- and cross-polarized E_{12} excitations in the circumferential dimension. Z-polarized states are more localized and their transition energies are strongly stabilized as tube curvature is reduced. In contrast, cross-polarized states are completely delocalized in the azimuthal dimension and their transition energies are less sensitive to the tube diameter. Therefore, the splitting displayed in Fig. 4.13 increases with tube diameter. Finally, we notice that the optical transition dipole moment of cross-polarized excitations is expected to be sensitive to perturbations that break the tube symmetry in the circumferential dimension, such as inhomogeneous electric fields, and varying local dielectric constants appearing due to tube bundling and defects. Such perturbations affect properties of cross-polarized excitations due to their complete delocalization in the radial dimension. Indeed, recent experiments report that the optical signatures of cross-polarized excitations are more pronounced in bundles as compared to individual tubes [Luo *et al.* (2008)].

In summary, due to degeneracy of the corresponding molecular orbitals, each considered transition between the van Hove singularities gives rise to four distinct excitonic bands. Excitations originating from E_{11} and E_{12}/E_{21} transitions have transition dipole moments parallel- and cross-polarized to the tube axis, respectively. However, most of these transitions are optically forbidden (dark). We typically observe a single, strongly optically allowed excitation related to the E_{11} transition and two near degenerate, weakly allowed excitations related to the E_{12} and E_{21} transitions. The transition energies of all dark excitons span the entire range from the lowest state to the E_{22} excitation. All considered excitons represent tightly bound electron-hole quasi-particles, which have very similar delocalization properties along the tube axis. Such properties driven by excitonic effects are dramatically different from predictions from one electron theory assigning cross-polarized transition to be exactly in the middle between E_{11} and E_{22} transitions [Wang *et al.* (2007)].

To examine excitonic properties and to classify the resultant transitions, we have analyzed the transition density matrices using the natural helical coordinate system. We found that photoexcited charge densities follow the tube helicity. Consequently, such a representation provides a natural universal framework for all SWNTs. Our analysis allows one to recognize critical distinctions between parallel- and cross-polarized transitions. In particular, all z-polarized excitons are more localized in the circumferential dimension as compared to the cross-polarized excitations. This difference is so well pronounced that it allows us to separate the four parallel-polarized E_{11} excitons from the cross-polarized E_{12} (E_{21}) transitions. Consequently, the energy splitting between parallel- and cross-polarized excitons increases with increasing tube diameter, since the energy of the optically allowed E_{11} exciton is more sensitive to the tube curvature. On the other hand, we expect the optical signatures of delocalized cross-polarized excitons to be very sensitive to the variation of the dielectric environment around the tube, as well as tube defects. These effects may be used in future applied spectroscopic measurements as a complementary characterization of nanotube samples. In addition, studies of the helical nature of charge distribution in SWCNTs brings us closer in understanding of self-assembling mechanisms of SWNTs functionalized by DNA strands [Yarotski *et al.* (2008, in press)]. Although DNA-wrapped SWNTs promise broad applications in metal-semiconductor tube separation and unbundlinng procedures[Zheng *et al.* (2003a)], drug delivery and cancer therapy[Kam *et al.* (2005)], very little is still known about details of SWNT-DNA hybrid formation and its combined properties. Numerical observation of charge and exciton distribution along a tube chirality is a first step towards rationalization of a strong binding of DNA to the tube surface, while preserving a helical configuration [Yarotski *et al.* (2008, in press)].

4.3.3 *Bright Excitons at Ground and Excited Geometries*

Now we focus on three bright parallel polarized excitons and on a question how the photoexcitation dynamics changes the exciton structure. The contour plots presented in the first and second column in Fig. 4.14 represent the transition density matrices between the ground and electronically excited E_{11} and E_{22} levels, respectively, in Cartesian coordinates. The rel-

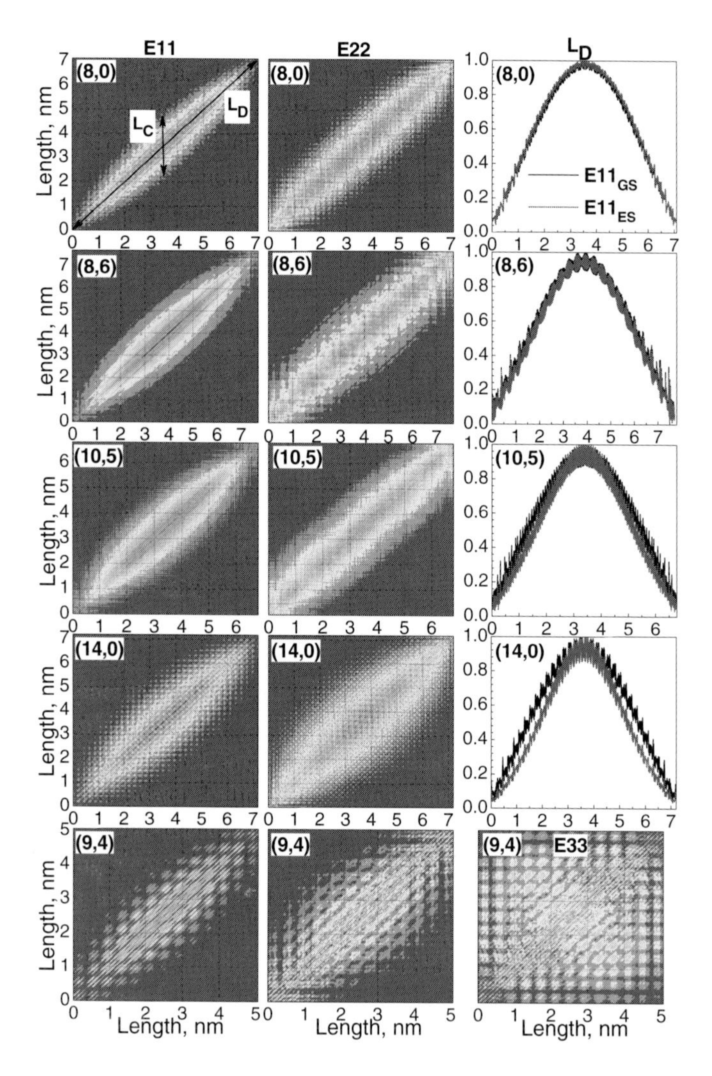

Fig. 4.14 Comparison of the first (E_{11}) and second (E_{22}) excitonic transition densities at ground (GS) and the lowest optically excited (ES) state optimal geometries for various tubes. The first and second columns display two-dimensional contour plots of transition density matrices for E_{11} and E_{22} excitons, respectively, as a function of electron (Y-axis, nm) and hole (X-axis, nm) coordinates at GS geometry. The bottom row also shows the third excitonic transition (E_{33}) calculated for a (9,4) tube, corresponding to delocalized excitation. Transition density magnitude scales from red (0) to violet (1) through the natural rainbow color sequence. The third column shows the diagonal slice (L_D) of E_{11} transition density matrix as a function of electron-hole pair position along a tube. Black and red lines correspond to GS and ES geometries, respectively. Tubes with larger diameter exhibit higher localization of the E_{11} exciton in the middle of a tube (self-trapping) at ES geometries compared to the GS profile. For color reference, turn to page 158.

evant cross-sections along the diagonal L_D are shown in the right column in Fig. 4.14, which compares localization of the exciton at the ground- and excited-state geometries. As expected, Fig. 4.14 demonstrates that the center of mass of E_{11} and E_{22} excitons is spread over the entire tube for all studied SWCNTs. Such exciton delocalization patterns are identical to that observed in conjugated polymers [Tretiak and Mukamel (2002); Tretiak *et al.* (2002, 2007b)]. There are a few other similarities between excitonic features in these molecular systems. For example, L_D plots for all studied tubes have periodically repeating peaks in the electronic density due to photoexcitation. The chiralities of SWCNTs determine the shape and the period of these peaks. These periodic variations are a signature of the spontaneous weak dimerization of the π electronic density [Tretiak *et al.* (2007b)]. Such effects are especially pronounced in the conjugated polymer case [Tretiak *et al.* (2002)]. The dimerization is very weak in the narrow tubes (e.g., (8,0)) because of bonding strain and disruption of π-conjugation induced by the tube curvature. The peak amplitudes increase in the medium-diameter tubes (e.g., (14,0)), reflecting enhanced mobility of π electronic systems. However, subsequent reduction is expected because of smaller $\pi - \sigma$ mixing and vanishing Peierls dimerization in super-wide SWCNTs and planar graphene sheet [Saito *et al.* (1998a)]. These electronic phenomena have clear vibrational signatures as well and lead to Peierls distortion, which will be discussed in the next subsection.

Another important characteristic of the electronic excitation is the exciton coherence size (maximal distance between electron and hole) given by the largest off-diagonal extent L_C of the non-zero matrix area. Contour plots in Fig. 4.14 show that L_C is finite for both E_{11} and E_{22} excitations for all tubes. This corresponds to the tightly bound singlet excitons, with exciton size being 3-5 nm (depending on the tube diameter, see discussion below). Other excitons are present in SWCNTs as well. For example, triplet excitations recently calculated by TDDFT approach [Tretiak (2007a)] are even stronger bound excitons with a L_C size of about 1 nm; These excitations are 0.2-0.3 eV below their singlet counterparts. Different types of singlet excitations, appearing in the nonlinear [Korovyanko *et al.* (2004a); Zhao *et al.* (2006)] and two-photon [Wang *et al.* (2005a); Maultzsch *et al.* (2005)] absorption spectra of SWCNTs, are weakly bound excitons [Tretiak (2007a); Zhao *et al.* (2006)]. Such excitations are typical and have been extensively studied in many other 1-D materials, such as conjugated polymers [Tretiak and Mukamel (2002); Tretiak *et al.* (2002); Zhao *et al.* (2006); Gierschner *et al.* (2007)].

The E_{33} state calculated for a (9,4) tube (bottom right panel in Fig. 4.14) is nearly uniformly delocalized over the entire tube, with a L_C size comparable to the tube length. The analogous delocalization of the E_{33} exciton is observed for (7,6) tube, as well. This corresponds to either an unbound or weakly bound excitonic state [Araujo *et al.* (2007)]. The interpretation of E_{33} being a delocalized excitation, unlike bound E_{11} and E_{22} states, agrees with experimental results and show that E_{33} and E_{44} optical transitions exhibit different diameter scaling compared to that of E_{11} and E_{22} [Araujo *et al.* (2007)]. For the E_{11}, E_{22}, and E_{33} excitations, Fig. 4.15 plots the distributions of the photoexcited electron wave functions when the hole has been fixed in the middle of the (7,6) tube, averaged over the radial distribution, e.g. L_C size. As illustrated by Fig. 4.15, the E_{11} and E_{22} states correspond to tightly bound excitons with a maximum electron-hole separation not exceeding 4 nm. This agrees well with previous theoretical

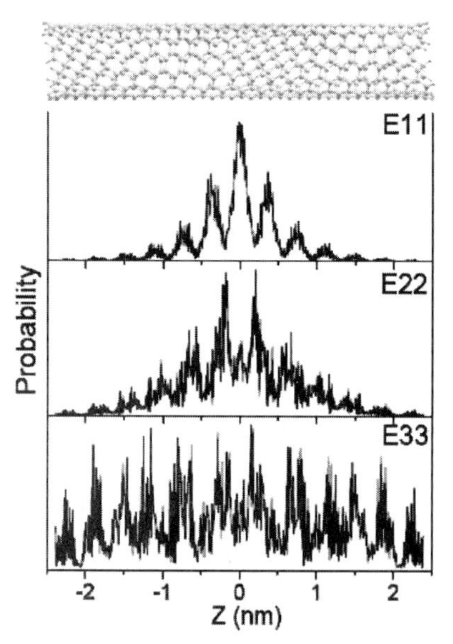

Fig. 4.15 Traces from top to bottom plot the electronic probability distribution for excited electrons on E_{11}, E_{22}, and E_{33}, respectively, along the (7,6) SWNT shown on top, considering the hole fixed in the middle.

studies of excitonic effects in SWNTs [Perebeinos *et al.* (2004); Chang *et al.* (2004); Zhao and Mazumdar (2004a)]. The E_{33} state, however, displays a very different behavior. It is nearly uniformly delocalized over the entire tube, confined by the tubes ends. This corresponds to either an unbound or a weakly bound excitonic state. Furthermore, note that while the calculations have been made for SWCNTs of ~ 0.88 nm in diameter, the same result should hold for wider diameters since the exciton binding energy is predicted to decrease with increasing diameter, scaling as $1/d$ [Perebeinos *et al.* (2004); Chang *et al.* (2004); Zhao and Mazumdar (2004a); Kane and Mele (2004)].

However, this contradicts the previous theoretical studies based on GW approximation [Spataru *et al.* (2004)], where E_{11}, E_{22} and E_{33} levels are all tightly bound excitons. Such a discrepancy may point to important electronic structural features of SWCNTs, such as the mixing and various non-Condon interactions between E_{33} states and the continuum of the lower lying E_{11} and E_{22} bands. Our calculations estimate less than 0.001 eV separation in the density of states at the E_{33} level attributed to other molecular states, compared to about 0.02 eV separation at the E_{22} level. Subsequently any small perturbation (e.g., temperature or dielectric inhomogeneity and disorder) may result in interactions and superpositions of delocalized and localized states in experimental samples. In finite tube calculations these perturbations come naturally from the end effects and imperfect optimal geometries, whereas the wavevector **k** allows for separation of 'pure' excitonic bands in the calculations with imposed periodic boundary conditions. We note that such non-Condon phenomena were recently observed at E_{22} level in SWCNTs [Manzoni *et al.* (2005b); Yin *et al.* (2007); Shreve *et al.* (2007a)].

Finally we discuss scaling of the exciton size L_C with the tube diameter. A recent study, which utilized a tight-binding method parameterized by *ab initio* (GW/LDA) results, proposes a simple scaling relation for L_C, having a leading dependence on diameter D with chirality corrections [Capaz *et al.* (2006)]. These calculations define exciton sizes as Gaussian width of the exciton wave function $|\psi|^2$ along the tube axis, and obtain the L_C extent of E_{11} in the range of 1-2 nm [Capaz *et al.* (2006)]. In another model calculation, the exciton size L_C was prescribed to the root mean square (rms) distance between electron and hole [Perebeinos *et al.* (2004)]. Obviously L_C can be quite arbitrary and subjectively defined as the exciton wave function span along the tube axis. We suggest that a more practical definition of L_C, in conjunction with experimental conditions, should include the 'full' extent of the wave function. We recall that in other 1D systems, such as conjugated polymers, a saturation of physical observables (e.g., transition energies and polarizabilities) is developed when the system becomes larger than the exciton size L_C. This constitutes the fundamental transition from

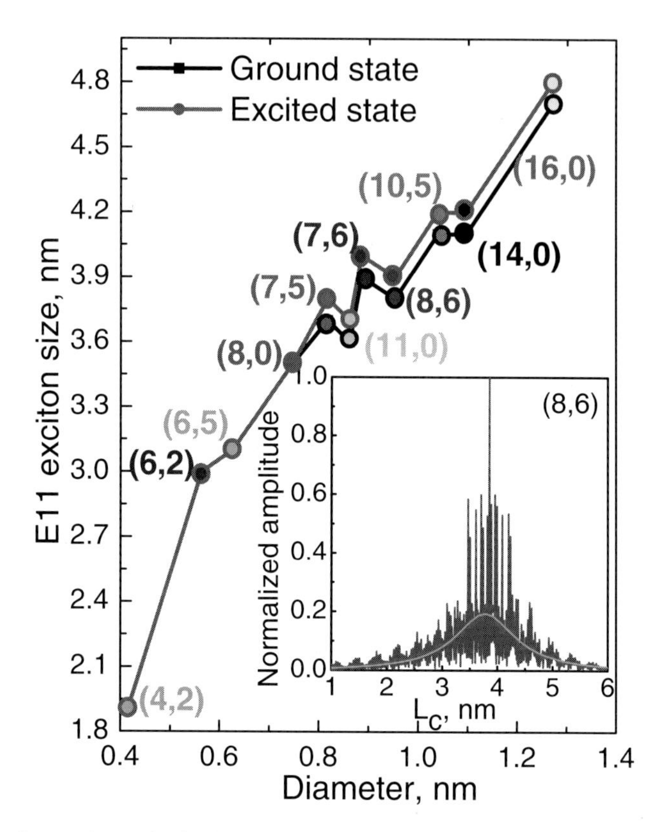

Fig. 4.16 Comparison of calculated E_{11} exciton size at GS (black line) and ES (red line) geometries for various tubes. The insert shows an example of the L_C slice of the transition density matrix of a (8,6) tube as a function of electron coordinate, when the hole is fixed in the middle of a tube (black line). The red line stands for the averaged transition density over the radius; the green line shows the corresponding Lorenzian fitting. The exciton size (maximal distance between an electron and a hole) is defined as the width of Lorenzian fit at 0.1 of its height, which approximately corresponds to a 90% drop of the excitonic wavefunction. The size of an exciton is increasing with tube diameter. The exciton size is slightly larger at the ES geometry for wide tubes, where vibrational relaxation leads to the smoother surface. For color reference, turn to page 159.

quantum confinement to 'bulk' regime [Samuel *et al.* (1994); Brédas *et al.* (1999); Tretiak and Mukamel (2002); Gierschner *et al.* (2007)], which occurs well beyond L_C sizes defines as the half-width of the distribution.

In our simulations we fitted the L_C cross-section of the E_{11} transition density matrices (Fig. 4.14), which corresponds to the excitonic wave function $|\psi|^2$ along the tube axis by a Lorenzian function (an example of the fit is shown in the inset in Fig. 4.16). The L_C exciton size is further approximated by the width of Lorenzian function at one tenth of its maximum amplitude. The resulting dependence of calculated L_C on the tube diameter (Fig. 4.16) spans a 2-5 nm range. Overall the curve roughly follows previously observed $\sim D$ dependence (due to better π-electron conjugation in wider SWCNTs), except for the very narrow (4,2) tube. Our absolute L_C magnitudes significantly overestimate previous calculations [Capaz *et al.* (2006)], because of the different definitions discussed above, but correlate well with commonly accepted values of about 2-3 nm determined for conjugated polymers [Brédas *et al.* (1999); Gierschner *et al.* (2007)]. The vibrational relaxation leading to ES geometries (discussed in the next subsection) results in a slight increase of the exciton size in large tubes (compare black and red lines in Fig. 4.16). In narrow tubes, this effect disappears and the exciton size stays almost unchanged upon the relaxation of the photoexcitation.

4.3.4 *Photoexcited Vibronic Dynamics: Peierls Distortions*

As expected, electron-phonon coupling leads to different optimal geometries of ground and excited states. Figure 4.17 shows comparison of GS and ES structures calculated for zigzag nanotubes by presenting the changes of a tube radius versus its azimuthal angle. Each data-point in Fig. 4.17 corresponds to the position of a carbon atom along a cross-section perpendicular to the tube axis and taken in the middle of a tube to reduce possible edge effects of finite SWCNTs. For all studied zigzag tubes, we observe two main maxima and minima in the radius dependence on the angle. This indicates a slightly elliptical shape of the cross-section for both GS and ES geometries, deviating from a perfect cylindrical geometry. We also note that chiral SWCNTs do not show such elliptical aberrations at both GS and ES geometries (not shown).

An important feature of the GS geometry is the corrugation of a tube surface [Tretiak *et al.* (2007b)] appearing as sharp peaks in the radial dependence in Fig. 4.17. Such surface corrugation in the GS geometry obviously makes some bonds shorter and some bonds longer in the hexagonal cell, which, in turn, reflects the related variations of the electronic density shown in the right column in Fig. 4.14. Similar to zigzag tubes chiral SWCNTs also have distorted GS geometry [Tretiak *et al.* (2007b)]. Due to rigid structure, the respective nanotube bond-length alternation is about 0.01-0.02 Å, which is at least ten times smaller than that in conjugated poly-

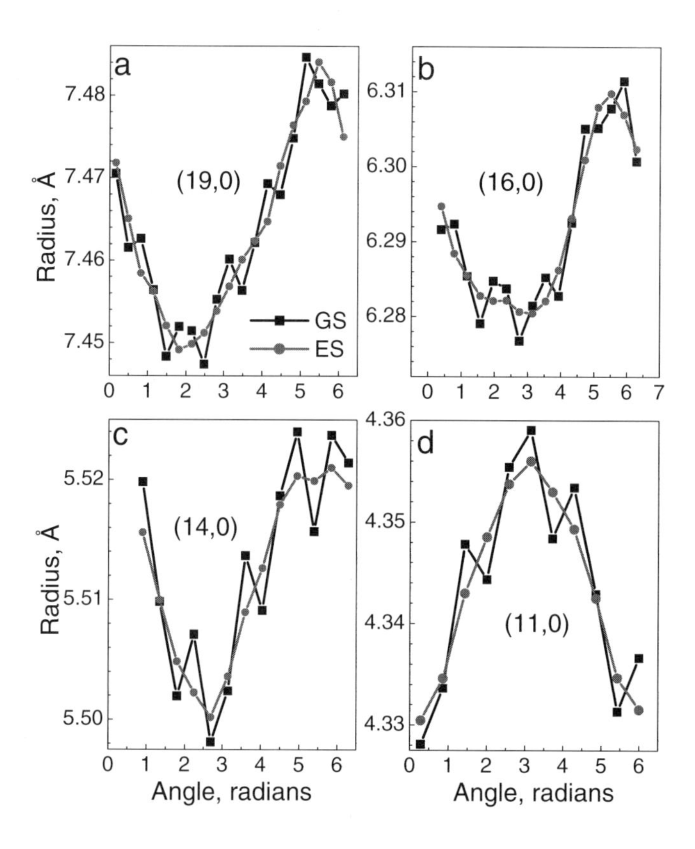

Fig. 4.17 Variation of tube radius as a function of azimuthal angle for GS (squares, black line) and ES (circles, red line) geometries of various zigzag tubes. GS geometry is corrugated for all considered tubes, while ES gemoetry exhibits smoother surface shape and, thus, decreased Pierels distortion. For color reference, turn to page 159.

mers. This geometric distortion, caused by interactions of $\pi - \sigma$-electron system due to curvature, has many common features with the effect of spontaneous symmetry breaking also known as Peierls distortion. The latter phenomenon is attributed to the instability of the Fermi energy states in quasi-one-dimensional metals. In half-filled zone materials, the instability is removed by electron-phonon interactions leading to gap opening at the Fermi level (metal-semiconductor transition), and induced by the folding of the Brillouin zone and related doubling of the real-space lattice period [Peierls (1955)]. Conjugated polymers provide an example of such transition with the gap opening of about 2 eV [Heeger *et al.* (1988)].

Peierls distortion has been extensively studied in metallic nanotubes [Saito *et al.* (1998a); Luis *et al.* (2006); Rafailov *et al.* (2005); Connetable *et al.* (2005); Bohnen *et al.* (2004a); Figge *et al.* (2001)]. In contrast to polymers, the produced band gap in the latter is smaller than the energy of thermal fluctuations. In the previous study [Tretiak *et al.* (2007b)], we considered semiconductor SWCNTs, and find that their surface experiences corrugation fully analogous to the lattice period doubling. Since this spontaneous symmetry breaking is associated with the electron-phonon interaction, we refer to it as Peierls distortion. However, this phenomenon can not be considered as metal-semiconductor transition.

In the ES geometry, the nanotube surface becomes locally less distorted. That is reflected by a smoother radial dependence on the angle (red lines in Fig. 4.17). Figure 4.18 shows the difference between the radial dis-

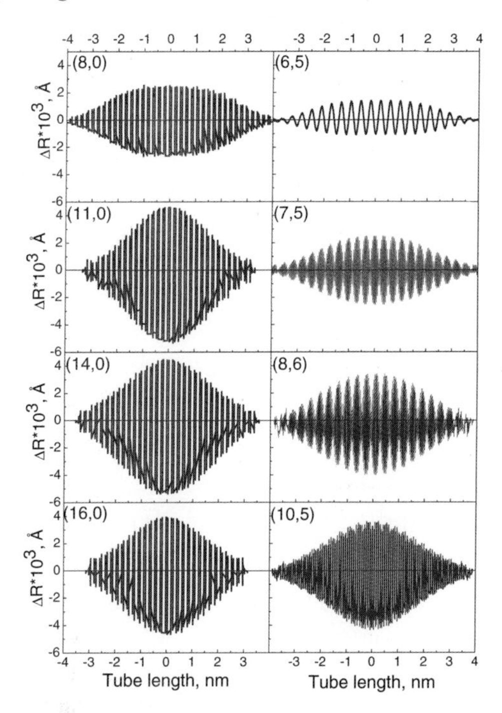

Fig. 4.18 Change of the radius $\Delta R = R_{ES} - R_{GS}$ with carbon coordinate along the tube Z axis for various tubes. Chiral and zigzag tubes are shown in the first and second columns, respectively. The diameters of tubes are increasing from the top to the bottom. Wider tubes have larger difference between GS and ES geometries due to substantial excitonic relaxation, which leads to self-trapping of exciton, shown in Fig. 4.14.

tance of carbon atoms along the tube length in the ground and excited geometries for both zigzag and chiral SWCNTs. The periodic oscillations reflect the decrease of Peierls distortions upon relaxation of photoexcitation. The largest geometric changes develop in the middle of the tubes. Overall, zigzag tubes show larger structural relaxation in the ES than chiral SWCNTs (compare left and right columns Fig. 4.18) in agreement with the previous work [Goupalov *et al.* (2006)]. This vibrational relaxation vanishes in the limit of very narrow tubes [Tretiak *et al.* (2007b)] due to disruption of π-conjugation and curvature induced strain. The radial difference in Fig. 4.18 shows that vibronic relaxation increases and reaches a maximum for tubes with diameters $\sim$1 nm. For tubes which are wider than 1 nm, the radial difference becomes smaller (compare (16,0) and (14,0) tubes in Fig. 4.18)), and approaches to zero in the limit of infinitely wide tubes. This shows the vanishing effect of tube curvature leading to the GS geometry distortion and related excited state vibronic relaxation.

The lattice distortions in the middle of the tube (Fig. 4.18) result in the localization of E_{11} exciton in the center of the molecule (compare black and right lines in the right column in Fig. 4.14), known as exciton self trapping in SWCNTs [Tretiak *et al.* (2007b)] and conjugated polymers [Tretiak *et al.* (2002)]. This can be clearly seen in both chiral [Tretiak *et al.* (2007b)], and particularly zigzag tubes with $\sim$ 1 nm diameter (e.g., (14,0) tube demonstrates the largest excitonic localization near the center as shown in Fig. 4.18). Subsequently, the related diagonal L_D size reduces upon vibrational cooling. In contrast, due to a smoother potential in the absence of Peierls distortions, the L_C coherence size slightly increases in the ES geometries (Fig. 4.16).

4.3.5 *Electron-Phonon Coupling: Huang-Rhys Factor, Stokes Shift, and Vibrational Relaxation Energies*

Nanotubes have been considered to be weakly exciton-phonon-coupled systems in the past, but recent PL excitation and photoconductivity results suggest that this may not be the case [Chou *et al.* (2005); Qiu *et al.* (2005)]. The quantitative determination of exciton-phonon coupling strength can be experimentally derived through the ratio of Raman fundamental to overtone intensities depends upon exciton-phonon coupling strength [Yin *et al.* (2007); Shreve *et al.* (2007a); Gambetta *et al.* (2006b)]. This modeling yields the Huang-Rhys factor (S), which is commonly used to characterize linear vibrational coupling to electronic excitations through the difference between ground and excited state geometries (See details in Subcection 4.2.2). In molecular spectroscopy, Franck-Condon overlap factors, related to S, determine the probability of specific electronic-vibrational transitions that give rise to vibrational replicas in absorption and fluorescence spectra. Subsequently, S characterizes the strength of exciton-vibration coupling, with values of $S \leq 0.1$ interpreted as weak exciton-vibration coupling.

Here we calculate Huang-Rhys factors for the E_{11} and E_{22} transitions in 5 different nanotube chiralities belonging to the $(n-m)mod3 = -1$ (referred to as "mod-2") group and also compare the calculated results with experimental Raman profiles. The results have been published re-

tube	ω_ν cm^{-1} (exp.)	ω_ν cm^{-1} (th.)	S E_{22} (exp.)	Δ E_{22} (exp.)	S_A E_{22} (th.)	Δ_A E_{22} (th.)	S_A E_{11} (th.)	Δ_A E_{11} (th.)	S E_{11} (th.)	Δ E_{11} (th.)
(10,2)	266	326	0.025	0.224	0.0085	0.130	0.0056	0.106	0.00684	0.117
(9,4)	257	315	0.013	0.161	0.0105	0.145	0.0069	0.118	0.00756	0.123
(8,6)	245	298	0.017	0.190	0.0163	0.180	0.0123	0.157	0.01232	0.157
(10,5)	226	274	0.017	0.190	0.0100	0.141	0.0076	0.123	0.00966	0.139

cently [Shreve *et al.* (2007a)], and summarized in Table 4.2. As can be seen from the data, Calculated displacements using 'exact' (optimal geometries) and 'approximate' (obtained by Eq. (4.6)) approaches for E_{11} agree within about 10% for RBM for selected nanotubes, which translates to about 20% differences in the Huang-Rhys factors. The deviation of 'approximate' Δ from the experimental one is ranging from 5% to 25% for (9,4), (8,6) and (10,5) tubes, which demonstrate pretty well agreement. Experimental and theoretical results deviate the most for the tube (10,2), due to its short length (4.5 nm) that introduce artificial edge effects to the calculations. From the analysis of data presented in Table 4.2, exciton-phonon coupling for the RBM is found to be in a weak coupling regime.

The same conclusion can be also derived from comparing the average diameter change upon E_{11} exciton relaxation, presented in Fig. 4.18 in the previous Subsection. Thus, relaxation of the first exciton leads either to slight tube buckling, mostly for mod-1 tube family, or narrowing for mod-2 tubes. This situation inverses for the E_{22} exciton case [Goupalov *et al.* (2006); Araujo *et al.* (2007)]. The average diameter change is, however, small compared to the bond-length changes [Tretiak *et al.* (2007b)]. Subsequently, the RBM has smaller electron-phonon coupling compared to that of longitudinal G-mode. Typical calculated dimensionless displacements along E_{11} potential energy surface for RBM and G-modes are 0.1-0.15 and 0.3-0.6, respectively, for SWCNTs with diameter $\sim$1 nm (detailed studies have been published recently [Yin *et al.* (2007); Shreve *et al.* (2007a); Gambetta *et al.* (2006b)]). Such values correspond to a weak electron-phonon coupling regime compared to most molecular materials. Nevertheless, local distortions of the tube surface during vibronic relaxation (Fig. 4.18) lead to experimentally detectable coherent phonon dynamics and anharmonic coupling between RBM and G-mode appearing in the excited state potential [Gambetta *et al.* (2006b)].

To characterize the effects of electron-phonon coupling on the electronic degrees of freedom, we further analyze the Stokes shift (defined as a difference between vertical transition energies at GS and ES geometries) and the energy change of the E_{11} excited state upon vibrational relaxation. Figures 4.19 and 4.20 show calculated values of these quantities as a function of tube length and diameter, respectively. The greater the ES geometries deviate from the equilibrium GS structure, the stronger the energy changes are, as evidently observed in Figs. 4.19 and 4.20. The shorter tubes exhibit stronger vibrational relaxation effects (Fig. 4.19). This is expected because molecular systems in a quantum confinement regime usually undergo larger change of the wave function upon electronic excitation, and, subsequently, stronger vibronic effects. Such trend, for example, was explored in conjugated oligomers [Tretiak *et al.* (2002)]).

When tubes are much longer than the corresponding exciton size L_C, both Stokes shift and vibrational relaxation energies evidence noticeable convergence to the relatively low value, comparable with thermal energy. Both quantities converge roughly as $1/L$ with deviations at large tube lengths. Such convergence is similar to what we observe for energy gap scaling (Fig. 4.6) and usage of fitting formulae like Eq. (4.9), may be necessary for extrapolation to the limiting values. Unfortunately, excited state relaxation in long tubes with wider than 1 nm diameter was not calculated due to large numerical expense of the deployed method.

Unlike generally observed $1/L$ or $1/D$ trends, the dependence of the Stokes shift and vibrational relaxation on tube diameter presented in Fig. 4.20, is strongly non-linear. Such dependence matches the above mentioned concept of diameter-dependent rigidity. The narrowest tube (4,2) with a diameter comparable to the phenyl ring size, does not show any

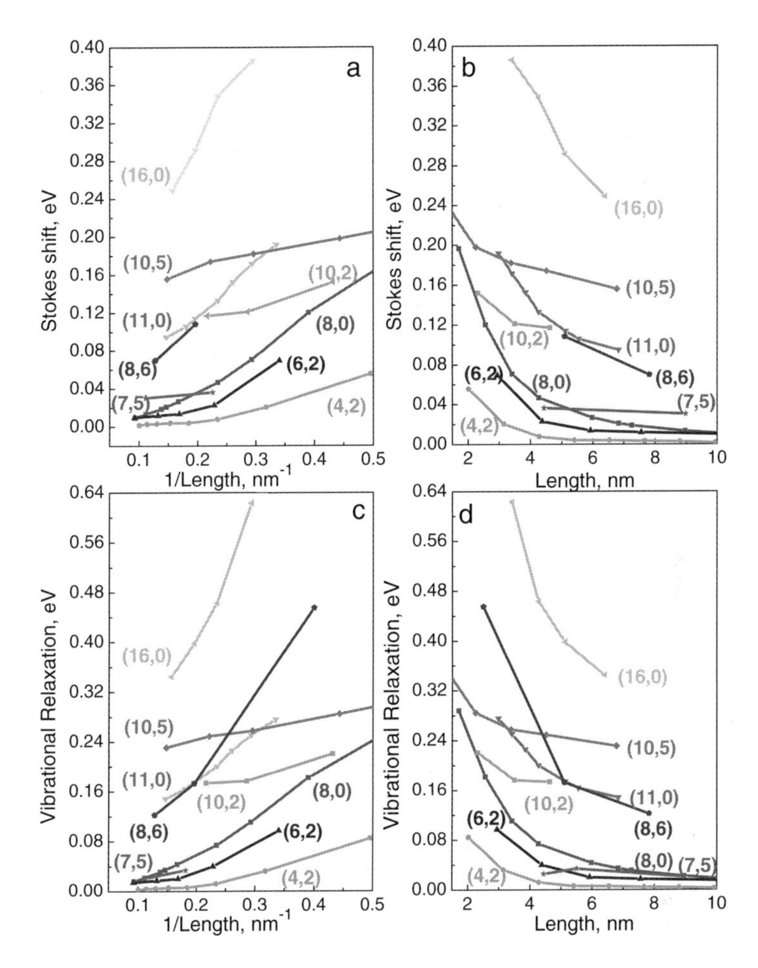

Fig. 4.19 Variation of calculated electron-phonon coupling with **(a)**, **(c)** inverse tube length and **(b)**, **(d)** tube length. The Stokes shift and excited state vibrational relaxation are shown in panels **(a)**-**(b)** and **(c)**-**(d)**, respectively.

vibrational effects due to disrupted π-conjugation and considerable bonding system strain, that does not allow carbon atoms to move upon optical excitation. Slightly larger tubes such as (6,2), (8,0), and (6,5) have very small relaxation. Tubes with diameter $\sim$ 1 nm are less rigid, have better π-electron mobility, and, consequently show $\sim 50 - 200$ meV Stokes shift and $\sim 30 - 90$ meV excited state relaxation. Such effects are detectable in the absorption/emission profiles of SWCNTs. Particularly, even though the main transition in the spectra corresponds to 0-0 line, weak vibrational replica (particularly G-mode) are clearly seen [Chou *et al.* (2005); Htoon *et al.* (2004)], which amounts to the 'spectral weight transfer' of about 10% to electron-phonon lines [Qiu *et al.* (2005); Plentz *et al.* (2005a); Tretiak *et al.* (2007b); Perebeinos *et al.* (2005a)].

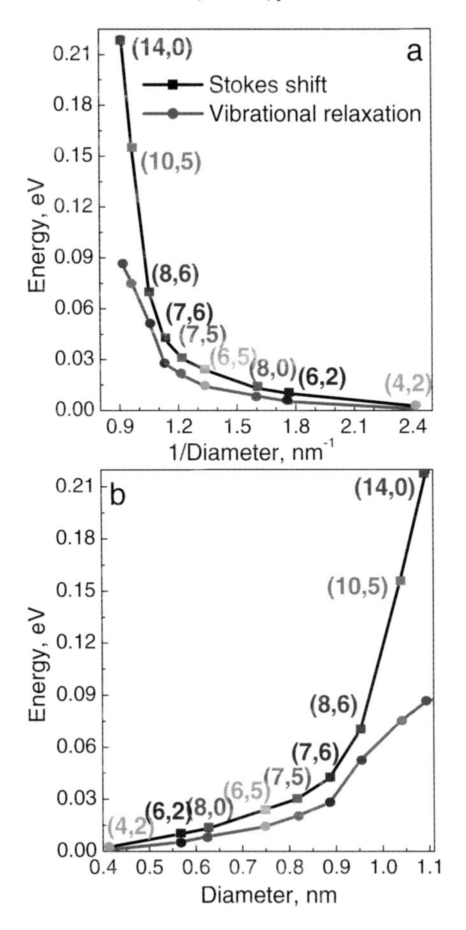

Fig. 4.20 Variation of calculated Stokes shift and excited state vibrational relaxation as a function of **(a)** inverse tube diameter and **(b)** tube diameter.

Finally, we note that Stokes shifts are usually defined in experiment as the spectral shift between absorption and emission maxima. In molecular systems with large vibrational coupling and displacements, this roughly coincides with the difference between vertical transition energies at GS and ES geometries, which is identical to computational definition of Stokes shift. Comparison of experimental and calculated Stokes shifts in SWCNTs is, however, not straightforward. Due to relatively small vibrational effects, the main transition in the spectra of SWCNTs corresponds to 0-0 line, consequently, very small shifts are observed between maxima of absorption and emission profiles. These shifts in the 0-0 line are related to the environmental relaxation and possible exciton transport along the tube. However, these quantities do not reflect internal vibrational relaxation of the tube, which is the origin of fairly substantial computed Stokes shifts (see Fig. 4.19 and 4.20). Subsequently, to compare calculations and experimental measurements, shifts of the center of mass of absorption and emission profiles (including complete vibrational progressions) need to be evaluated. These should provide experimental estimates and comparisons of internal vibrational relaxation and environmental relaxation in SWCNTs, and is a subject of future studies.

4.4 Conclusions

Investigations of electronic structure and spectroscopic properties of SWCNTs are of primary importance for successful development of novel electronic and photonic devices based on SWCNTs, as well as for ongoing global progress in nanoscience and nanotechnology. Recent realization of the significance of electron-electron interactions and electron-phonon coupling in SWCNTs immediately emphasized the usefulness of modern numerically expensive quantum-chemical calculations [Spataru *et al.* (2004); Chang *et al.* (2004)], compared to model-type analytical approaches [Saito *et al.* (1998a)]. In addition to the solid-state based approaches assuming homogeneous/ordered infinite SWCNTs [Spataru *et al.* (2004); Zhao and Mazumdar (2004a); Perebeinos *et al.* (2004); Chang *et al.* (2004)], finite size calculations have addressed many important aspects of nanotube photophysics [Zhao and Mazumdar (2004a); Zhao *et al.* (2006); Zhou *et al.* (2004); Gambetta *et al.* (2006b); Tretiak *et al.* (2007b); Tretiak (2007a)]. The last approach is legitimate if the lengths of the considered tube segments are larger than the corresponding diameter and the exciton size L_C.

Our reported ESMD simulations of finite SWCNTs provide important details of photoinduced dynamics, that were not accessible previously. In addition to the optimal geometries and single-point electronic structure, which are available from traditional simulations, our approach directly explores excited state vibrational relaxation and exciton-phonon coupling, accounting for curvature effects and related $\sigma - \pi$ interactions. Good agreement with experimental data is achieved for energies of excitonic transitions (e.g., Fig. 4.6 and [Tretiak *et al.* (2007b)]), electron-phonon coupling constants [Shreve *et al.* (2007a)], and time-resolved experimental data [Gambetta *et al.* (2006b)]. The fundamental effects, such as tightly bound and unbound excitons [Araujo *et al.* (2007)], Peierls-like distortions [Tretiak *et al.* (2007b)], and exciton self-trapping [Tretiak *et al.* (2007b); Plentz *et al.* (2005b)], are found to be important features of excited state structure in SWCNTs. Substantial excited state displacements and Huang-Rhys factors [Yin *et al.* (2007); Shreve *et al.* (2007a)] computed in the tubes with ~ 1 nm diameter, results in the noticeable Stokes shifts and vibrational relaxation energies, which might be examined experimentally [Yin *et al.* (2007); Shreve *et al.* (2007a)]. Our calculations quantitatively explore the scaling dependence of excitonic and vibrational properties of SWCNTs on their diameters, lengths and chiralities. In particular, the explored length-dependence, besides being a tool for extrapolating data to the limiting values, may have broader physical content.

Solid-state approaches assuming uniform infinite tube lengths characterize excitons with L_C size, i.e. by spatial separation between photoexcited electron and hole. Molecular type finite-size calculations necessarily introduce L_D size related to the position of the center-of-mass of the exciton along the tube. On the ideal infinite tubes, of course, this quantity should have a trivial uniform distribution. However, in the experimental samples the tubes are subject to dielectric environment, solvent induced disorder, local defects, chemical functionalizations, intertube interactions, etc. All these effects can lead to the spatial localization of the exciton (e.g., Anderson-type localization). Such phenomena are difficult to model using infinite-tube approach with imposed periodic boundary conditions. In contrast, all these perturbations and inhomogeneities are routinely accessible within molecular approaches dealing with finite tube segments. Intelligent manipulation of nanotubes ensembles and their assembly into functional devices, which make use of emergent properties, are aims of current technological and synthetic work. Understanding the role of possible defects, disorder effects and intertube interactions present theoretical challenges, where the molecular-type approaches could make significant contributions in the future studies.

Chapter 5

Carbon Nanotube Technological Implementations

As the research and applications of nanotechnology continue forward, CNTs consistently establish themselves as a key building block of the future. A major reason for this is that SWCNTs can be thought of as near perfect self-assembled macromolecules. Pristine SWCNTs have very low defect concentrations and high length to diameter aspect ratios [Fan *et al.* (2005)], which lends them to a wide variety of applications. Many of their unique properties are derived from their carbonic nature. Their composition of pure carbon inhibits defect formation, especially compared to other nanotubes derived from several elements, such as BN or BNC nanotubes. Due to carbon's sp^2 hybridization and the single atomic layer that forms SWCNTs, there are no dangling bonds to pacify and every atom is on the surface [Avouris *et al.* (2007)]. This has important consequences for their mechanical and electronic properties. SWCNTs have been envisioned in applications ranging from molecular electronics to biological sensors and energy production and storage. We will now give a short review of some recent demonstrations of the potential of SWCNTs, beginning with molecular electronics and field effect transistors (FET).

Molecular electronics is the next step in making things smaller and faster. Not just integrated circuits, either, as creating and understanding nanoscale architectures has important implications for photovoltaics, batteries, sensors, and diodes. The ultimate goal is to create integrated circuits using single molecules and connecting them with molecular wires. This allows for smaller power consumption and faster switching times. It also requires a true revolution in synthesis and characterization techniques, as well as a deep understanding of the fundamental properties of nanoscale systems.

CNTs have drawn a large amount of interest in the molecular electronics research field. CNT FETs (CNFET) were proven in 1998 [Tans *et al.* (1998)] and a large variety of devices have been demonstrated since. A FET is a three terminal device that is at the heart of modern computer processors. In this architecture, a current flows between two electrodes, known as the source and drain. This current is modulated by the third electrode, known as the gate, allowing the creatation of ON and OFF states. The rate of switching between these ON and OFF states determines the speed of the device. Much of the interest is motivated by the vibrational spectrum of CNTs. At low excitation energies or bias, electrons only weakly couple to the acoustic modes. The energies of optical photons are near 180 meV, providing a large energy window where there is little scattering of the electrons during transport [Avouris and Chen (2006)]. It has been shown that electrons in CNTs can travel ballistically (without scattering) for distances of up to tens of μm. Little to no scattering in devices prevents energy from being lost to heat, as well as allowing switching speeds at gigahertz frequencies. These characteristics are particularily important for FETs and quantum wires. The mobility of electrons in CNFETs was measured at over 100,000 cm^2/Vs, breaking the previous record of approximately 77,000 cm^2/Vs in InSb [Durkop *et al.* (2004)]. This is also over 100 times greater than that of silicon.

Current FETs have dimensions that are much larger than the molecular electronics of the future. Thus, a single SWCNT is too small to transport enough current for detection in a FET [Guo *et al.* (2005)]. As true molecular electronics are still far in the future, work as been undertaken to use arrays of aligned SWCNTs to replace the silicon channel in FETs [Guo *et al.* (2005)]. Arrays of aligned CNTs can carry enough current to be detected in contemporary, while taking advantage of nanotube switching speeds. However, there are also substantial hurdles to be cleared in creating better contacts between the SWCNTs and the electrodes. The mismatch between the work functions of SWCNTs and metal contacts creates large Schottky barriers, hindering performance and applications.

Of course, this does not mean that CNFETs have not already found practical applications. It has been shown that CNFETs can detect biological molecules at picomolar concentrations. The energy levels in SWCNTs and, hence, the current flowing through the FETs are very sensitive to environmental perturbations. The adsorption of tens of large biological molecules on to a FET is enough to produce a detectable change in current [Allen *et al.* (2007)]. This method takes advantage of the fact that in CNFETs the channel material is exposed to the outside environment.

In current Si based FETs, the channel material is covered by an insulating material. Further, as SWCNTs can be grown to typically micrometer lengths, they have good size compatiblity with biological molecules such as DNA and proteins. Glucose, DNA, and protein detection have been demonstrated using CNFETs, while functionalized CNFETs are able to detect a single virus [Wang (2005)]. Refinement of device design implies detection of single molecules, and, potentially, direct real-time observation of electronic profiles and dynamics in biological systems.

It has also been shown that SWCNTs can act as ideal diodes [Lee (2005)]. Interactions between the nanotube and the substrate create states within the SWCNT band gap and cause deviations from ideality. However, by etching about the substrate below a SWCNT channel, it was shown to perform as an ideal diode [Lee (2005)]. This has important applications in photovoltaic applications where the separation of electrons and holes create a photo-induced current. A potential problem in incorporating CNTs into photovoltaic devices is the rate of internal conversion, or phonon-induced relaxation of excited charge carriers. Relaxation of the excited electron and hole to the VB and CB edges results in energy loss to heat, as well as strongly localized excitons. Intelligent engineering of device (i.e. stacking of multiple diodes based on different diameter SWCNTs with different absorption spectra) is required to maximize solar efficiency. Estimates of the efficiency of a well-design SWCNT device are greater than 5%. It has further been shown that the junction between CNTs of different diameters creates a natural diode. This is due to the inverse relationship between a CNT's band gap and its diameter [Saito *et al.* (1998b)]. Although significant improvements in synthetic techniques are required to take advantage of this unique property, the ability to create a unipolar diode without patterning electrodes could greatly simplify the construction of SWCNT circuits.

A major difficulty in implementing CNFETs is the large variety of SWCNTs created during synthesis of these nanostructures. All synthetic methods produce mixtures of semiconducting and metallic SWCNTs, as well as a range of diameters with varying band gaps. Methods have been developed to eliminate metallic SWCNTs from ensembles of tubes, as well as synthesis aimed at narrowing the diameter distribution [Hertel *et al.* (2006)]. An interesting approach involves SWCNT 'seeding', where a single SWCNT is sliced into pieces. It has been shown that these seeds may be used to grow SWCNTs of the same chirality as the original piece [Smalley *et al.* (2006)]. Still, great advances in CNT synthesis will be required to enable large scale production of SWCNT devices. The isolation of graphene sheets has produced an alternative material to SWCNTs for use as the channel material in FETs. Large graphene sheets are semi-metals, and the lack of a band gap prohibits switching behaviour necessary for a FET [Avouris *et al.* (2007)]. However, theoretical and experimental efforts have shown that quantum confinement effects in narrow graphene ribbons open a band gap and could possibly provide a more readily available material for future molecular electronics.

Another area of interest is the development of transparent, conductive films incorporated with CNTs. The primary goal is to replace the expensive indium-tin oxide (ITO) layer used in may chromatic displays and prototype photovoltaic cells. It has been shown that SWCNTs included in polymeric blends increases the conductance. Low filling factors of the CNTs allows for good transmittance through the sheets. The CNTs are often functionalized to allow covalent bonding and in one study the CNTs were shortened to provide better interactions with the substrate [Jung *et al.* (2007)] . Further motivation for SWCNT-polymer blends is flexibility. ITO is brittle and inhibits flexible displays or solution processing of photovoltaic cells. Transparent, conductive polymer based films do not have this restriction [Lee *et al.* (2008)].

The energy storage applications of CNTs have begun to receive a large amount of attention as well. It has become obvious that petroleum based internal combustion engines are not sustainable (if only due to the large wealth and power imbued upon nations lucky enough to have these resources). Solar technology has received great interest as a replacement to generating electricity via fossil fuels. Another interesting proposal is using catalysts and solar rays to split water into hydrogen and oxygen. While CNTs are not directly envisioned for usage in these applications, work has been undertaken to utilize them in other aspects, such as hydrogen and electric storage.

Ultracapacitors are expected to challenge lithium ion batteries as the energy storage device of choice in the next decade. Contemporary super- or ultracapacitors lag far behind lithium-ion batteries in their ability to store energy density. However, batteries are slow to charge and wearout after several hundred cycles due to the electrochemical energy storage mechanism. Ultracapacitors based on SWCNTs take advantage of the high surface area to store charge [Frackowiak and Beguin (2001)]. As the energy is stowed as a separation of charges, it requires no moving parts and may be released and replenished much more quickly than batteries. Capacitors also store a much greater power density, which allows the electricity to be released in large amounts. This is an important for advancement for electric automobiles that often have difficulty accelerating quickly or climbing hills.

Hydrogen storage is also of great interest as van der Waals interactions between H_2 and the inner and outer walls of SWCNTs allows the molecules to absorb on the tubes. As carbon is a low elemental weight atom, SWCNTs potentially can be used to meet the Department of Energy's standards for hydrogen storage. Reports of the absorption and weight percent using SWCNTs vary greatly [Liu and Cheng (2005)] at present. A possible difficulty is the weakness of the van der Waals interactions between the tube and hydrogen. It has been estimated near the energy of vibrational motion of the carbon atoms at room temperature, discouraging adsorption. Lower temperatures have been shown to increase adsorption and chemical modification of the SWCNT sidewalls has been proposed. Research into absorption/desorption catalysts and optimization of the pressure and temperature conditions are need to create an efficient storage mechanism.

Carbon nanotubes have shown great promise for the advancement of many existing technological applications and for the development of novel devices and architectures. The simplicity of their structure coupled with intelligent design of nanodevices, such as FETs, diodes and sensors, has revealed them as a pivotal component of the nanotechnology revolution- all of this occurring within two decades of their revival. Further, the impressive mechanical properties of CNTs, though beyond the scope of this book, have inspired a great number of other novel applications [Salvetat *et al.* (2006)]. Most CNT devices are still in a proof-of-concept stage of development, and extensive challenges to their implementation as commercial devices still remain. It will likely take a decade or two before CNT devices are extensively integrated into common experimental and household electronics and materials. Regardless, it is a very exciting time for anyone in the nanotechnology field, especially those investigating nanostructured carbon.

Color Index

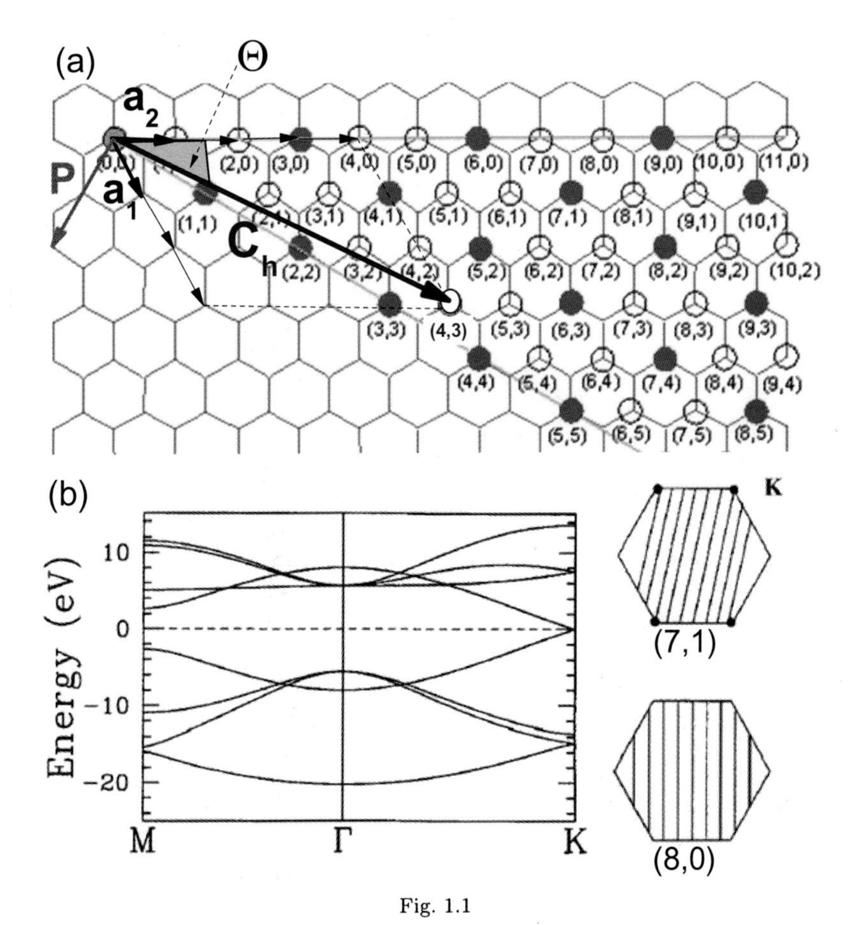

(a)
Θ
a_2
P
a_1
C_h
(0,0)
(2,0)
(3,0)
(4,0)
(5,0)
(6,0)
(7,0)
(8,0)
(9,0)
(10,0)
(11,0)
(1,1)
(3,1)
(4,1)
(5,1)
(6,1)
(7,1)
(8,1)
(9,1)
(10,1)
(2,2)
(3,2)
(4,2)
(5,2)
(6,2)
(7,2)
(8,2)
(9,2)
(10,2)
(3,3)
(4,3)
(5,3)
(6,3)
(7,3)
(8,3)
(9,3)
(4,4)
(5,4)
(6,4)
(7,4)
(8,4)
(9,4)
(5,5)
(6,5)
(7,5)
(8,5)
(b)
Energy (eV)
10
0
−10
−20
M
Γ
K
K
(7,1)
(8,0)

Fig. 1.1

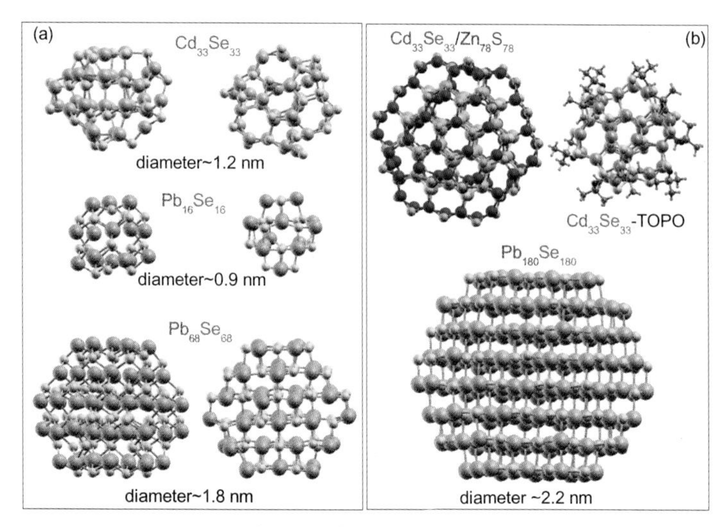
(a)
$Cd_{33}Se_{33}$
diameter~1.2 nm
$Pb_{16}Se_{16}$
diameter~0.9 nm
$Pb_{68}Se_{68}$
diameter~1.8 nm
$Cd_{33}Se_{33}/Zn_{78}S_{78}$
(b)
$Cd_{33}Se_{33}$-TOPO
$Pb_{180}Se_{180}$
diameter ~2.2 nm

Fig. 2.3

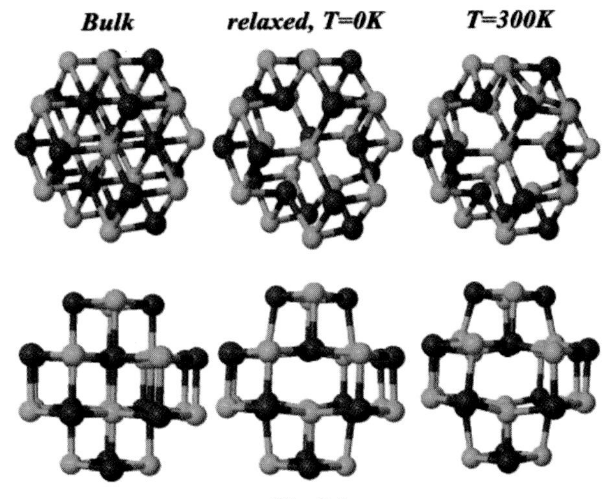
Bulk
relaxed, T=0K
T=300K

Fig. 2.4

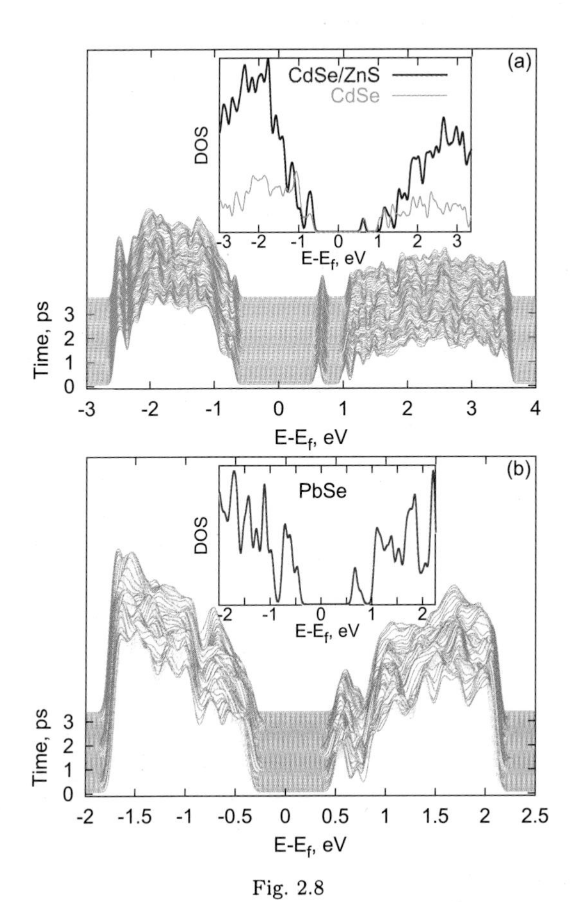
(a)
CdSe/ZnS
CdSe
DOS
E-E$_f$, eV
Time, ps
E-E$_f$, eV
(b)
PbSe
DOS
E-E$_f$, eV
Time, ps
E-E$_f$, eV

Fig. 2.8

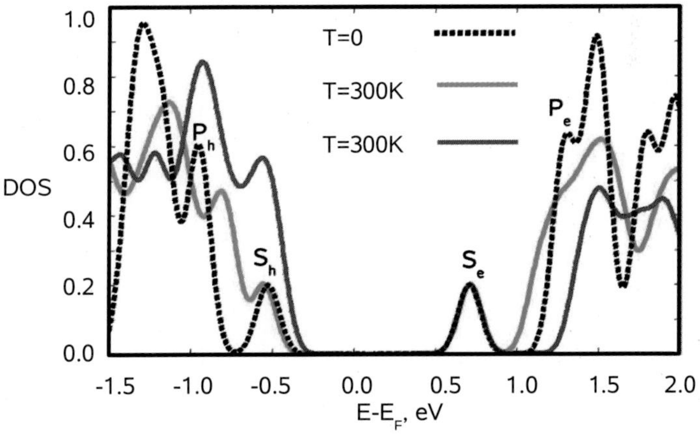

T=0
T=300K
T=300K
DOS
P$_h$
S$_h$
S$_e$
P$_e$
E-E$_F$, eV

Fig. 2.9

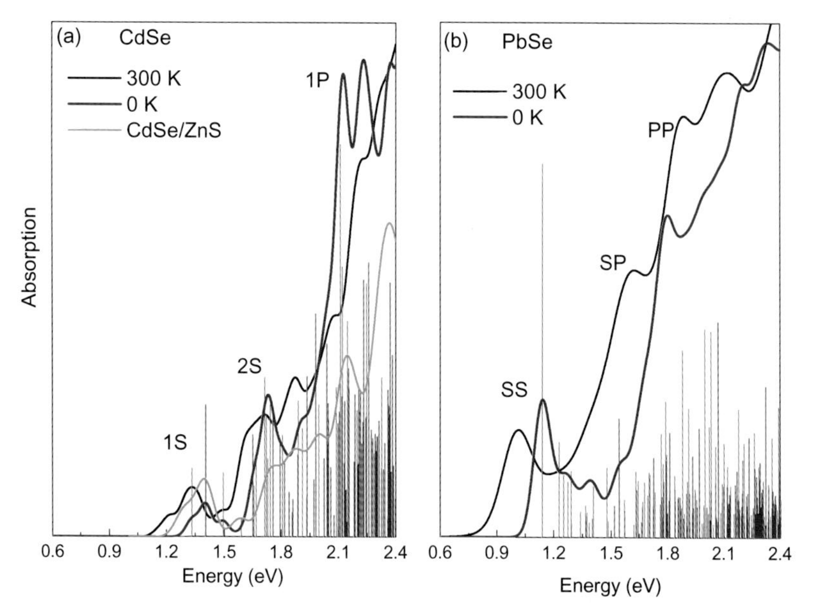

Fig. 2.10

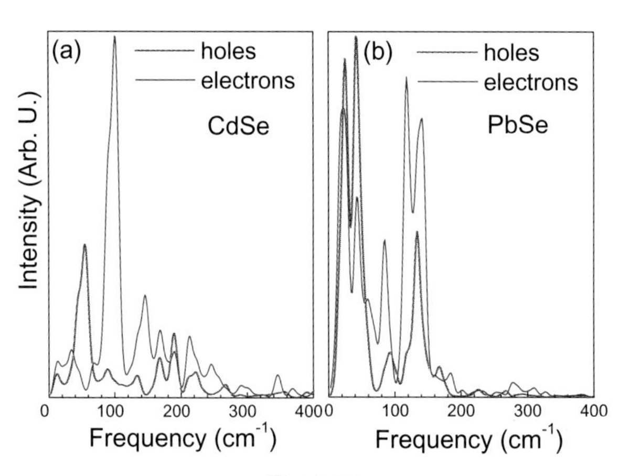

Fig. 2.12

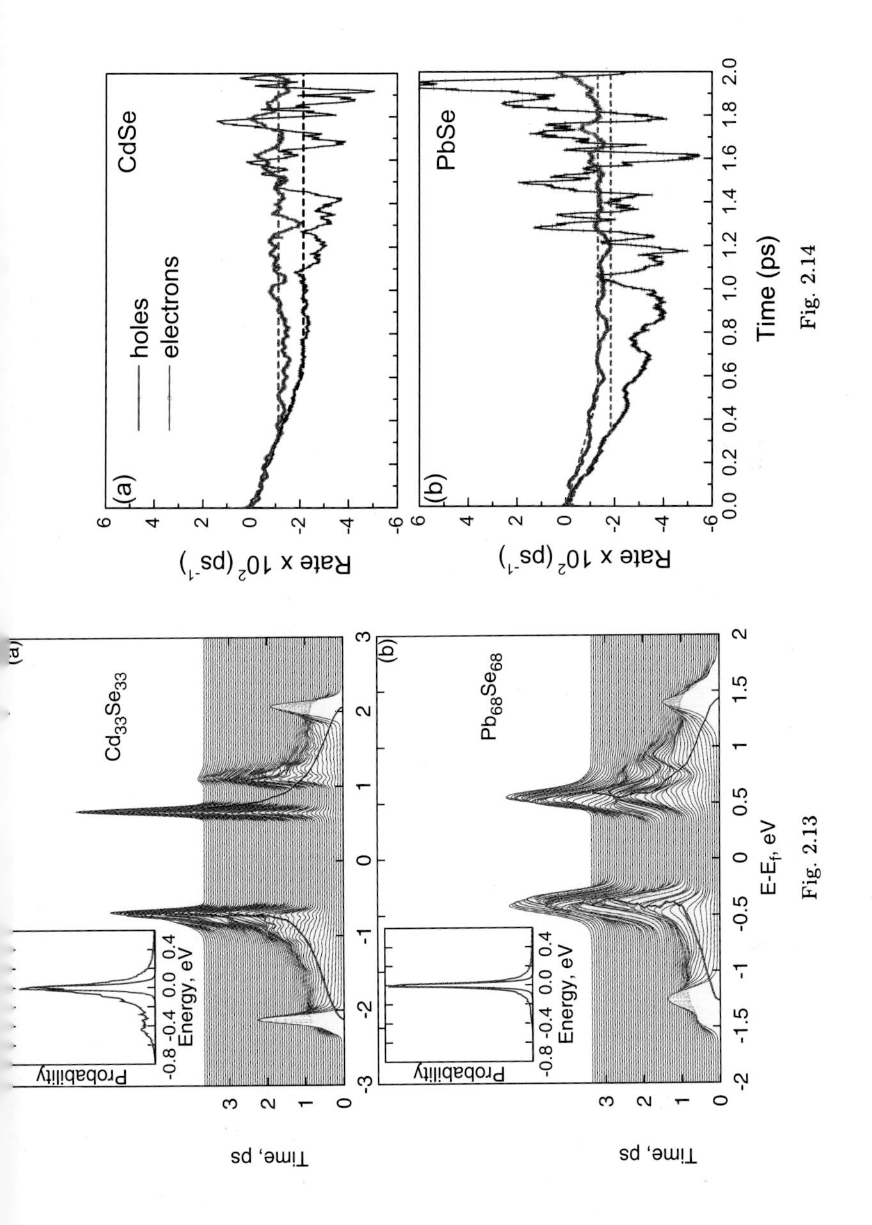

Fig. 2.13

Fig. 2.14

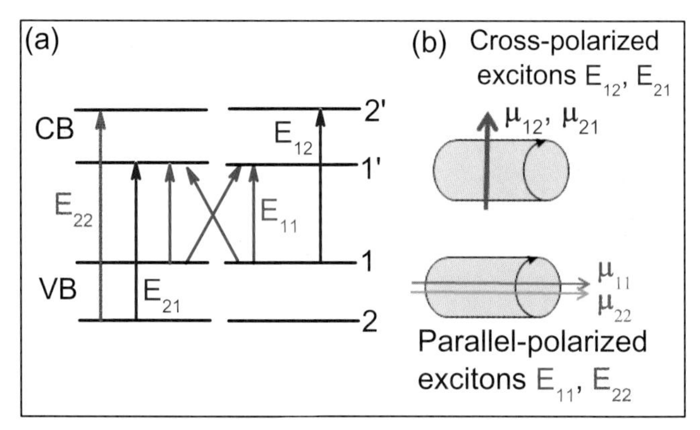
(a)
CB
VB
E_{22}
E_{21}
E_{12}
E_{11}
2'
1'
1
2
(b) Cross-polarized excitons E_{12}, E_{21}
μ_{12}, μ_{21}
μ_{11}
μ_{22}
Parallel-polarized excitons E_{11}, E_{22}

Fig. 4.1

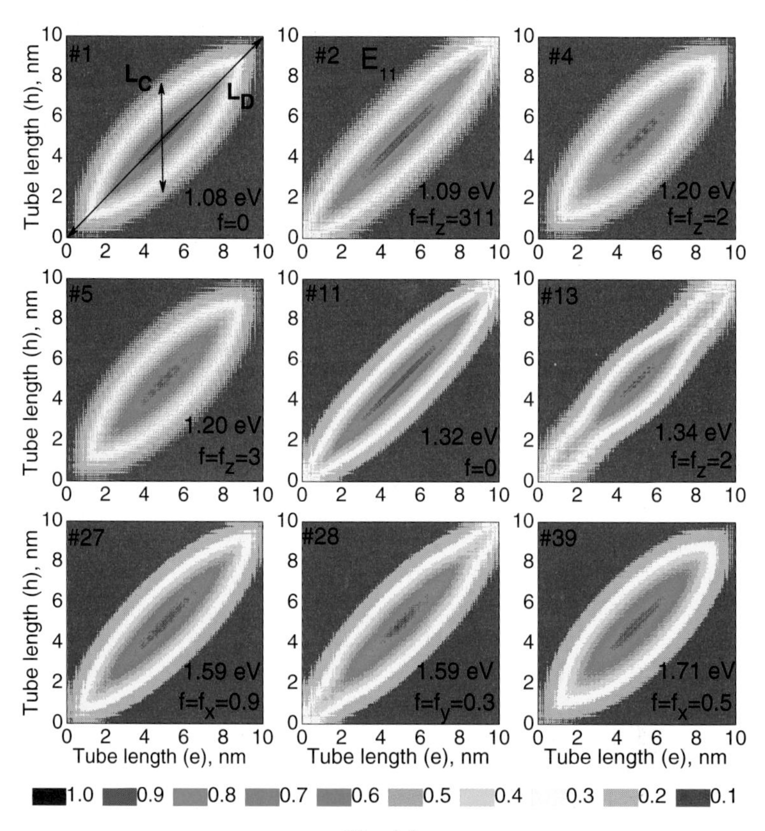
#1
L_C
L_D
1.08 eV
f=0
#2 E_{11}
1.09 eV
$f=f_z=311$
#4
1.20 eV
$f=f_z=2$
#5
1.20 eV
$f=f_z=3$
#11
1.32 eV
f=0
#13
1.34 eV
$f=f_z=2$
#27
1.59 eV
$f=f_x=0.9$
#28
1.59 eV
$f=f_y=0.3$
#39
1.71 eV
$f=f_x=0.5$
Tube length (h), nm
Tube length (e), nm
1.0 0.9 0.8 0.7 0.6 0.5 0.4 0.3 0.2 0.1

Fig. 4.8

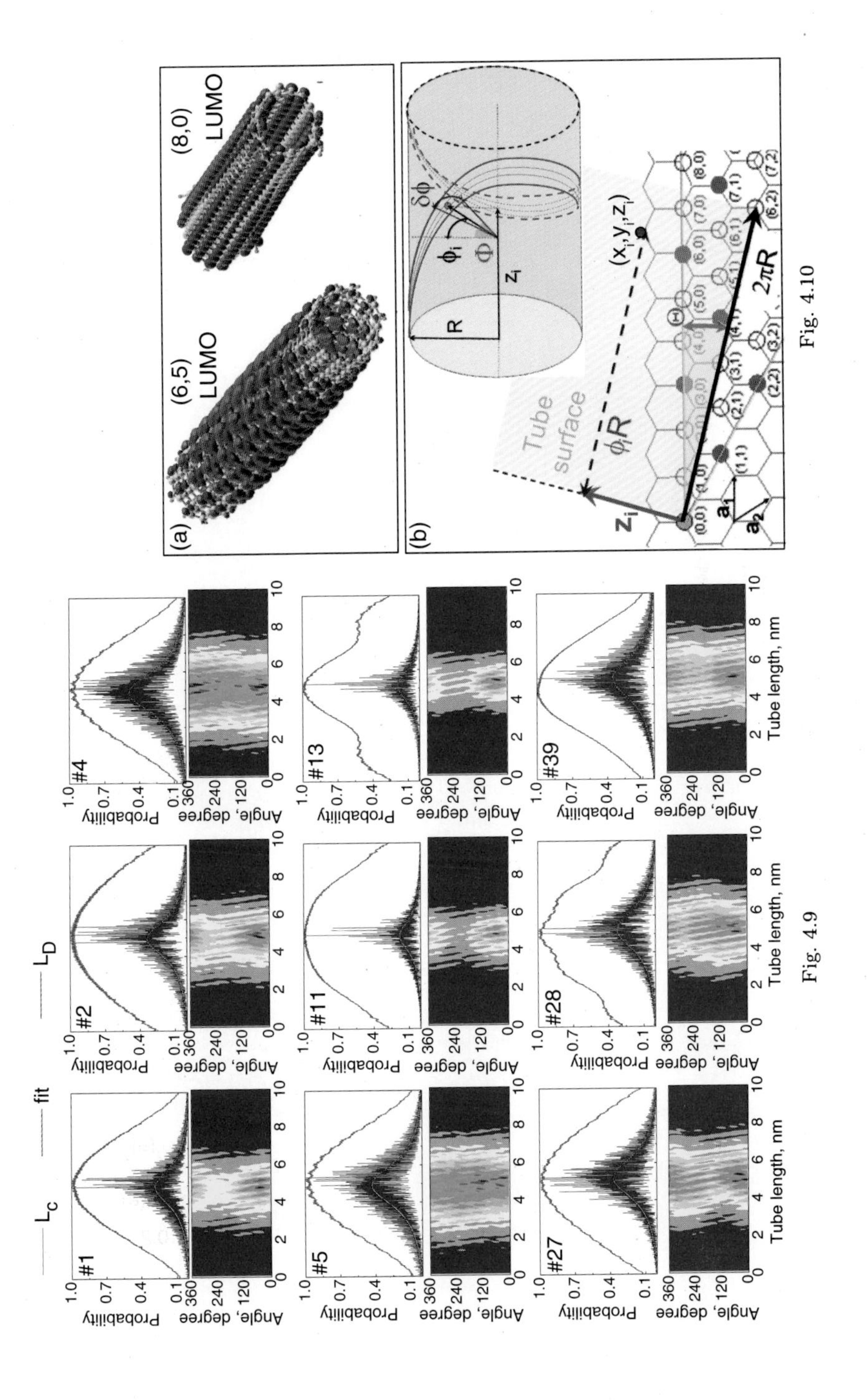

Fig. 4.9

Fig. 4.10

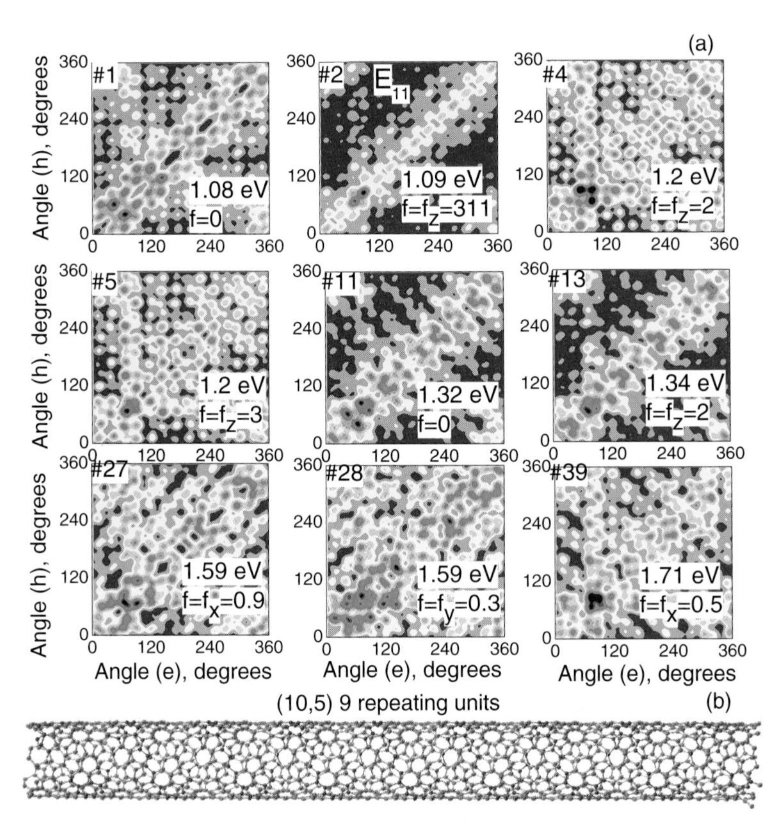
(a)
#1
1.08 eV
f=0
#2
E_{11}
1.09 eV
$f=f_z=311$
#4
1.2 eV
$f=f_z=2$
#5
1.2 eV
$f=f_z=3$
#11
1.32 eV
f=0
#13
1.34 eV
$f=f_z=2$
#27
1.59 eV
$f=f_x=0.9$
#28
1.59 eV
$f=f_y=0.3$
#39
1.71 eV
$f=f_x=0.5$
Angle (h), degrees
Angle (e), degrees
0
120
240
360
(10,5) 9 repeating units
(b)

Fig. 4.11

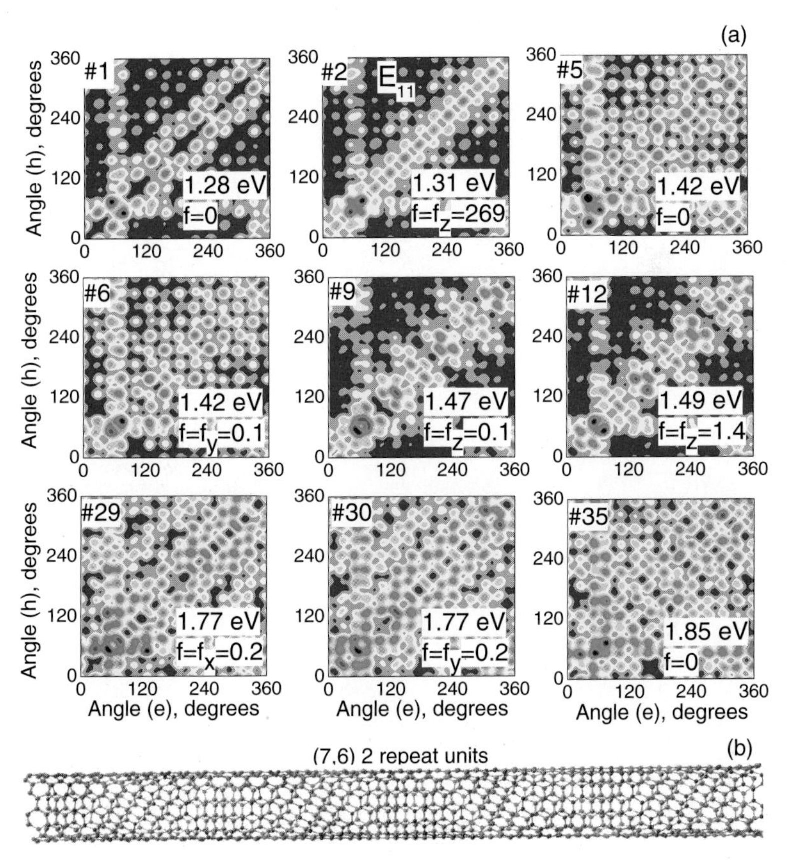
(a)
#1
1.28 eV
f=0
#2
E_{11}
1.31 eV
$f=f_z=269$
#5
1.42 eV
f=0
#6
1.42 eV
$f=f_y=0.1$
#9
1.47 eV
$f=f_z=0.1$
#12
1.49 eV
$f=f_z=1.4$
#29
1.77 eV
$f=f_x=0.2$
#30
1.77 eV
$f=f_y=0.2$
#35
1.85 eV
f=0
Angle (h), degrees
Angle (e), degrees
0
120
240
360
(7,6) 2 repeat units
(b)

Fig. 4.12

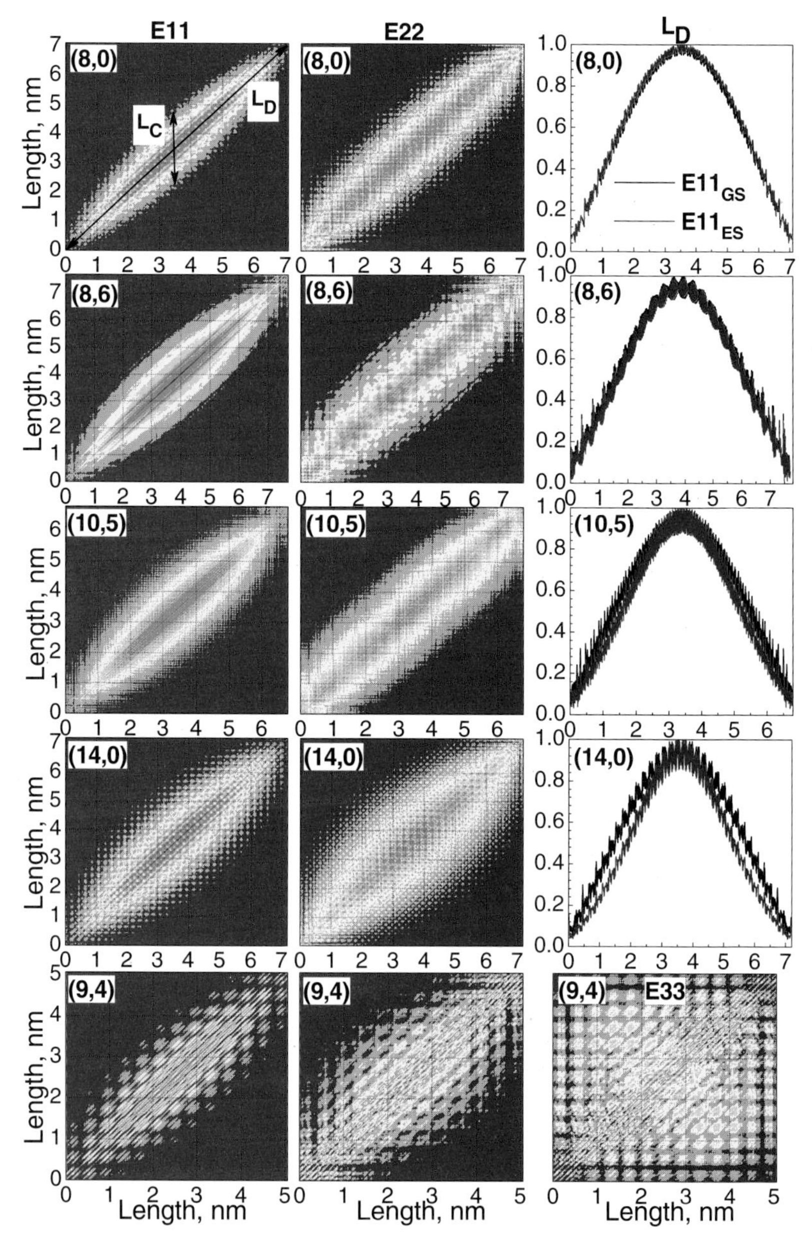
E11
E22
L_D
(8,0)
L_C
L_D
$E11_{GS}$
$E11_{ES}$
(8,6)
(10,5)
(14,0)
(9,4)
E33
Length, nm
Length, nm

Fig. 4.14

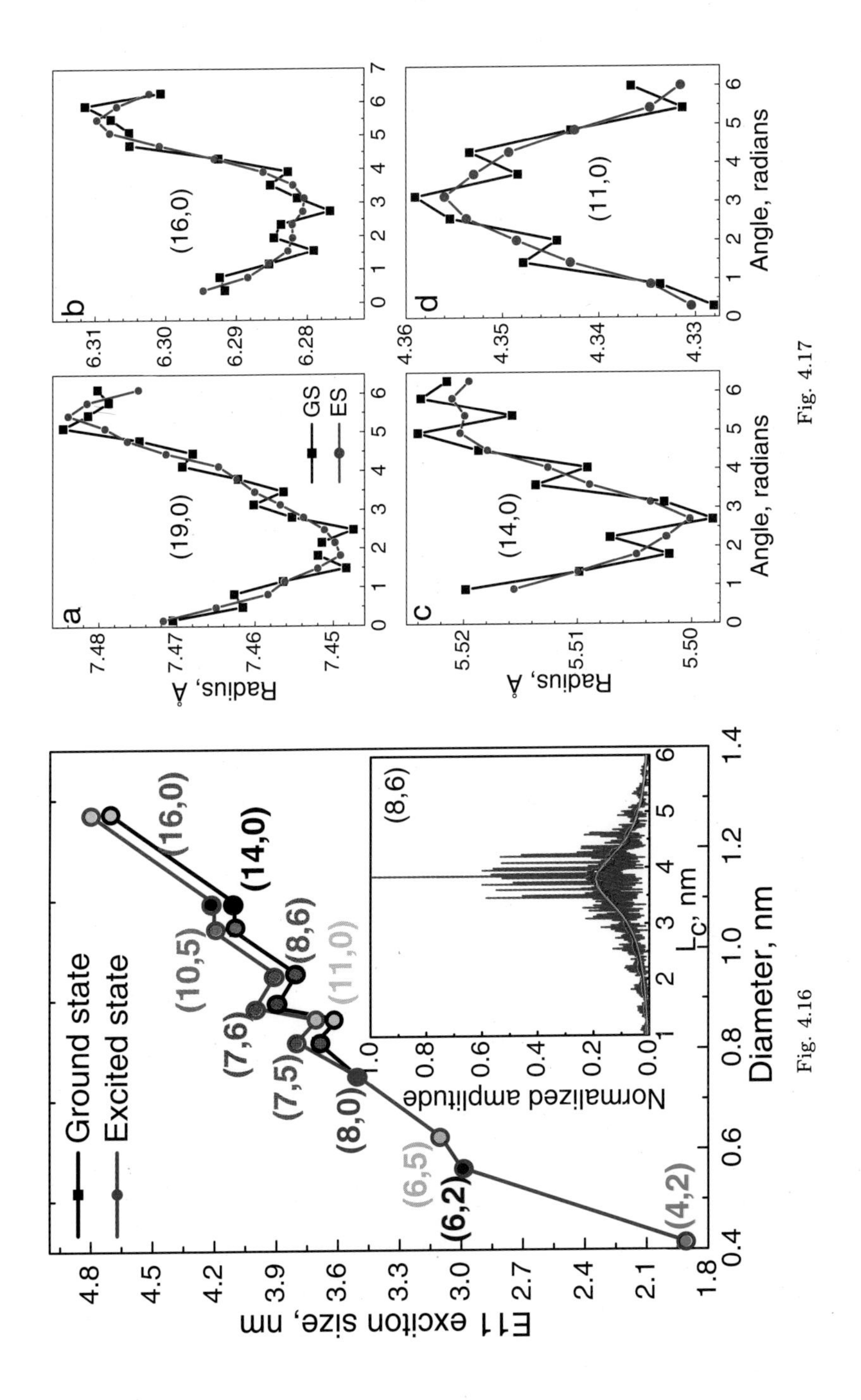

Fig. 4.16

Fig. 4.17

Bibliography

Achermann, M., Hollingsworth, J. A. and Klimov, V. I. (2003). Multiexcitons confined within a subexcitonic volume: spectroscopic and dynamical signatures of neutral and charged biexcitons in ultrasmall semiconductor nanocrystals, *Phys. Rev. B* **68**, p. 245302.

Adler, F., Geiger, M., Bauknecht, A., Haase, D. and Ernst, P. (1998). *J. Appl. Phys.* **83**, p. 1631.

Ajiki, H. and Ando, T. (1994). Aharonov-Bohm effect in carbon nanotubes. *Physica. B, Condensed matter* **201**, pp. 349 – 52.

Albanesi, E. A., Blanca, E. L. P. Y. and Petukhov, A. G. (2005). Calculated optical spectra of iv-vi semiconductors pbs, pbse and pbte, *Comput. Mat. Sci.* **32**, p. 85.

Allan, G. and Delerue, C. (2004). Confinement effects in pbse quantum wells and nanocrystals, *Phys. Rev. B* **70**, p. 245321.

Allen, B. L., Kichambare, P. D. and Star, A. (2007). Carbon nanotube field-effect-transistor-based biosensors, *Advanced Mater.* **19**, p. 1439.

Allen, M. P. and Tildesley, D. J. (1987). *Computer Simulation of Liquids* (Clarendon Press, Oxford).

An, J. M., Franceschetti, A., Dudiy, S. V. and Zunger, A. (2006). The peculiar electronic structure of pbse quantum dots. *Nano Lett.* **6**, 12, p. 2728.

Ando, T. (1997). Excitons in carbon nanotubes. *J. Phys. Soc. Jpn.* **66**, 4, pp. 1066 – 1073.

Andreev, A. D. and Lipovskii, A. A. (1999). Anisotropy-induced optical transitions in pbse and pbs spherical quantum dots. *Phys. Rev. B* **59**, 23, pp. 15402 – 4.

Araujo, P. T., Doorn, S. K., Kilina, S., Tretiak, S., Einarsson, E., Maruyama, S., Chacham, H., Pimenta, M. A. and Jorio, A. (2007). Third and fourth optical transitions in semiconducting carbon nanotubes, *Phys. Rev. Lett.* **98**, 6, 067401.

Avouris, P. and Chen, J. (2006). Nanotube electronics and optoelectronics, *Materials Today* **9**, p. 46.

Avouris, P., Chen, J. and Perebeinos, V. (2007). Nanotubes, *Nature Nanotech.* **2**, p. 605.

Bachilo, S. M., Strano, M. S., Kittrell, C., Hauge, R. H., Smalley, R. E. and Weisman, R. B. (2002). Structure-assigned optical spectra of single-walled carbon nanotubes. *Science* **298**, 5602, pp. 2361 – 2366.

Baer, R. and Neuhauser, D. (2004). Real-time linear response for time-dependent density-functional theory, *J. Chem. Phys.* **121**, p. 9803.

Baer, R. and Neuhauser, D. (2005). Density functional theory with correct long-range asymptotic behavior, *Phys. Rev. Lett.* **94**, p. 043002.

Baker, J. D. and Zerner, M. C. (1991). Characterization of the random phase approximation with the intermediate neglect of differential-overlap hamiltonian for electronic spectroscopy, *J. Phys. Chem.* **95**, 22, pp. 8614–8619.

Bakueva, L., Gorelikov, I., Musikhin, S., Zhao, X. S., Sargent, E. H. and Kumacheva, E. (2004). Pbs quantum dots with stable efficient luminescence in the near-ir spectral range. *Adv. mat.* **16**, 11, pp. 926 – 929.

Balzani, V. and Scandola, F. (1991). *Supramolecular Photochemistry* (Ellis Harwood, New York).

Banin, U., Lee, C. J., Guzelian, A. A., Kadavanich, A. V., Alivisatos, A. P., Jaskolski, W., Bryant, G. W., Efros, A. L. and Rosen, M. (1998). Size-dependent electronic level structure of inas nanocrystal quantum dots: Test of multiband effective mass theory. *J. Chem. Phys.* **109**, 6, pp. 2306 – 9.

Banyai, L., Gilliot, P., Hu, Y. Z. and Koch, S. W. (1992). Surface-polarization instabilities of electron-hole pairs in semiconductor quantum dots. *Phys. Rev. B* **45**, 24, pp. 14136 – 42.

Barone, V., Peralta, J. E. and Scuseria, G. E. (2005a). Optical transitions in metallic single-walled carbon nanotubes. *Nano Lett.* **5**, 9, pp. 1830 – 3.

Barone, V., Peralta, J. E., Wert, M., Heyd, J. and Scuseria, G. E. (2005b). Density functional theory study of optical transitions in semiconducting single-walled carbon nanotubes. *Nano Lett.* **5**, 8, pp. 1621 – 4.

Batista, E. R. and Martin, R. L. (2005). Exciton localization in a pt-acetylide complex, *J. Phys. Chem. A* **109**, 43, pp. 9856–9859.

Berger, S., Voisin, C., Cassabois, G., Delalande, C., Roussignol, P. and Marie, X. (2007). Temperature dependence of exciton recombination in single-walled carbon nanotubes, *Nano. Lett.* **7**, p. 398.

Biercuk, M. J., Mason, N. and Marcus, C. M. (2004). Local gating of carbon nanotubes. *Nano Lett.* **4**, 1, pp. 1 – 4.

Bohnen, K. P., Heid, R., Liu, H. J. and Chan, C. T. (2004a). Lattice dynamics and electron-phonon interaction in (3,3) carbon nanotubes. *Phys. Rev. Lett.* **93**, 24, pp. 245501 – 4.

Bohnen, K. P., Heid, R., Liu, H. J. and Chan, C. T. (2004b). Lattice dynamics and electron-phonon interaction in (3,3) carbon nanotubes, *Phys. Rev. Lett.* **93**, p. 245501.

Boudreaux, D. S., Williams, F. and Nozik, A. J. (1980). *J. Appl. Phys.* **51**, p. 2158.

Brédas, J. L., Cornil, J., Beljonne, D., dos Santos, D. A. and Shuai, Z. (1999). Excited-state electronic structure of conjugated oligomers and polymers: A quantum-chemical approach to optical phenomena, *Acc. Chem. Res.* **32**, 3, pp. 267–276.

Brus, L. E. (1984). Electron-electron and electron-hole interactions in small semiconductor crystallites: the size dependence of the lowest excited electronic state. *J. Chem. Phys.* **80**, p. 4403.

Califano, M., Franceschetti, A. and Zunger, A. (2007). Lifetime and polarization of the radiative decay of excitons, biexcitons, and trions in cdse nanocrystal quantum dots. *Phys. Rev. B* **75**, 11, pp. 115401 – 1.

Califano, M., Zunger, A. and Franceschetti, A. (2004). Efficient inverse auger recombination at threshold in cdse nanocrystals. *Nano Lett.* **4**, p. 525.

Capaz, R. B., Spataru, C. D., Ismail-Beigi, S. and Louie, S. G. (2006). Diameter and chirality dependence of exciton properties in carbon nanotubes. *Phys. Rev. B* **74**, 12, pp. 121401 – 1.

Casey, J., S.Bachilo and Weisman, R. (2008, in press). Efficient photosensitized energy transfer and near-ir fluorescence from porphyrin-swnt complexes, *J. Materials Chem.* .

Casida, M. E. (1995). *Recent Advances in Density-Functional Methods, Part I*, Vol. 3 (World Scientific, Singapore).

Champagne, B., Bulat, F. A., Yang, W. T., Bonness, S. and Kirtman, B. (2006). Density functional theory investigation of the polarizability and second hyperpolarizability of polydiacetylene and polybutatriene chains: treatment of exact exchange and role of correlation. *J. Chem. Phys.* **125**, 19, pp. 194114 – 1.

Chang, E., Bussi, G., Ruini, A. and Molinari, E. (2004). Excitons in carbon nanotubes: An ab initio symmetry-based approach, *Phys. Rev. Lett.* **92**, 19, p. 196401.

Chang, R., Hsu, J. H., Fann, W. S., Liang, K. K., Chiang, C. H., Hayashi, M., Yu, J., Lin, S. H., Chang, E. C., Chuang, K. R. and Chen, S. A. (2000). Experimental and theoretical investigations of absorption and emission spectra of the light-emitting polymer meh-ppv in solution. *Chem. Phys. Lett.* **317**, 1/2, pp. 142 – 52.

Chattopadhyay, D., Galeska, L. and Papadimitrakopoulos, F. (2003). A route for bulk separation of semiconducting from metallic single-wall carbon nanotubes, *J. Am. Chem. Soc.* **125**, 11, pp. 3370 – 3375.

Chen, Y. C., Raravikar, N. R., Schadler, L. S., Ajayan, P. M., Zhao, Y. P., Lu, T. M., Wang, G. C. and Zhang, X. C. (2002). Ultrafast optical switching properties of single-wall carbon nanotube polymer composites at 1.55 mm. *Appl. Phys. Lett.* **81**, 6, pp. 975 – 7.

Chen, Z., Du, X., Du, M.-H., Rancken, C. D., Cheng, H.-P. and Rinzler, A. G. (2003). Bulk separative enrichment in metallic or semiconducting single-walled carbon nanotubes. *Nano Lett.* **3**, 9, pp. 1245 – 9.

Chen, Z. H., Appenzeller, J., Lin, Y. M., Sippel-Oakley, J., Rinzler, A. G., Tang, J. Y., Wind, S. J., Solomon, P. M. and Avouris, P. (2006). An integrated logic circuit assembled on a single carbon nanotube. *Science* **311**, 5768, pp. 1735 – 1735.

Chernyak, V., Volkov, S. N. and Mukamel, S. (2001a). Electronic structure-factor, density matrices, and electron energy loss spectroscopy of conjugated oligomers, *J. Phys. Chem. A* **105**, 10, pp. 1988 – 2004.

Chernyak, V., Volkov, S. N. and Mukamel, S. (2001b). Exciton coherence and electron energy loss spectroscopy of conjugated molecules, *Phys. Rev. Lett.* **86**, 6, pp. 995–998.

Chestnoy, N., Hull, R. and Brus, L. E. (1986). Higher excited electronic states in clusters of znse, cdse, and zns: spin-orbit, vibronic, and relaxation, phenomena. *J. Chem. Phys.* **85**, 4, pp. 2237 – 42.

Chou, S. G., Plentz, F., Jiang, J., Saito, R., Nezich, D., Ribeiro, H. B., Jorio, A., Pimenta, M. A., Samsonidze, G. G., Santos, A. P., Zheng, M., Onoa, G. B., Semke, E. D., Dresselhaus, G. and Dresselhaus, M. S. (2005). Phonon-assisted excitonic recombination channels observed in dna-wrapped carbon nanotubes using photoluminescence spectroscopy. *Phys. Rev. Lett.* **94**, 12, pp. 127402 – 4.

Coe, S., Woo, W. K., Bawendi, M. and Bulovic, V. (2002). Electroluminescence from single monolayers of nanocrystals in molecular organic devices. *Nature* **420**, p. 800.

Cognet, L., Tsyboulski, D. A., Rocha, J. D. R., Doyle, C. D., Tour, J. M. and Weisman, R. B. (2007). Stepwise quenching of exciton fluorescence in carbon nanotubes by single-molecule reactions, *Science* **316**, p. 1465.

Coker, D. F. (1993). *Computer Simulations in Chemical Physics* (Kluwer Academic Publishers, Netherlands), pp. 315–377.

Connetable, D., Rignanese, G. M., Charlier, J. C. and Blase, X. (2005). Room temperature peierls distortion in small diameter nanotubes. *Phys. Rev. Lett.* **94**, 1, pp. 015503 – 4.

Craig, C. F., Duncan, W. R. and Prezhdo, O. V. (2005). Trajectory surface hopping in the time-dependent kohn-sham approach for electron-nuclear dynamics. *Phys. Rev. Lett.* **95**, 16, p. 163001.

Crooker, S. A., Hollingsworth, J. A., Tretiak, S. and Klimov, V. I. (2002). Spectrally resolved dynamics of energy transfer in quantum-dot assemblies: towards engineered energy flows in artificial materials. *Phys. Rev. Lett.* **89**, 18, p. 186802.

Dahan, M., Levi, S., Luccardini, C., Rostaing, P., Riveau, B. and Triller, A. (2003). Diffusion dynamics of glycine receptors revealed by single-quantum dot tracking. *Science* **302**, p. 442.

Davidson, E. R. (1975). Iterative calculation of a few of lowest eigenvalues and corresponding eigenvectors of large real-symmetric matrices, *J. Comp. Phys.* **17**, 1, pp. 87–94.

Davidson, E. R. (1976). *Reduced Density Matrices in Quantum Chemistry* (Academic Press, New York).

Dewar, M. J. S., Zoebisch, E. G., Healy, E. F. and Stewart, J. J. P. (1985). Am1: A new general purpose quantum mechanical molecular model. *J. Am. Chem. Soc.* **107**, 13, pp. 3902 – 3909.

Dexheimer, S. L., VanPelt, A. D., Brozik, J. A. and Swanson, B. I. (2000). Femtosecond vibrational dynamics of self-trapping in a quasi-one-dimensional system. *Phys. Rev. Lett.* **84**, 19, p. 4425.

Doorn, S. K., Heller, D. A., Barone, P. W., Usrey, M. L. and Strano, M. S. (2004). Resonant Raman excitation profiles of individually dispersed single walled carbon nanotubes in solution, *Appl. Phys. A* **78**, 8, pp. 1147 – 1155.

Dresselhaus, M. S. (2004). Applied physics: Nanotube antennas, *Nature* **432**, 7020, pp. 959 – 960.

Dresselhaus, M. S., Dresselhaus, G., Saito, R. and A.Jorio (2007). Exciton photophysics of carbon nanotubes, *Ann. Rev. Phys. Chem.* **58**, p. 719.

Dresselhaus, M. S. and Eklund, P. C. (2000). Phonons in carbon nanotubes. *Adv. Phys.* **49**, 6, pp. 705 – 814.

Dubay, O., Kresse, G. and Kuzmany, H. (2002). Phonon softening in metallic nanotubes by a peierls-like mechanism. *Phys. Rev. Lett.* **88**, 23, pp. 235506 – 4.

Duncan, W. R. and Prezhdo, O. V. (2007). Theoretical studies of photoinduced electron transfer in dye-sensitized tio2. *Ann. Rev. Phys. Chem.* **58**, pp. 143 – 184.

Durkop, T., Getty, S. A., Cobas, E. and Fuhrer, M. S. (2004). carbon nanotubes as quantum wires, *Nano Lett.* **4**, p. 35.

Efros, A. L. and Efros, A. L. (1982). Interband absorption of light in a semiconductor sphere. *Sov. Phys. Semicond.* **16**, 7, pp. 772 – 775.

Efros, A. L. and Rosen, M. (1997). Random telegraph signal in the photoluminescence intensity of a single quantum dot. *Phys. Rev. Lett.* **78**, 6, p. 1110.

Efros, A. L. and Rosen, M. (2000). Electronic structure of semiconductor nanocrystals. *Ann. Rev. Mat. Sci.* **30**, p. 475.

Ekimov, A. I., Efros, A. L. and Onushenko, A. A. (1985). unknown, *Solid State Commun.* **56**, p. 921.

Ekimov, A. I., Hache, F., Schanne-Klein, M. C., Ricard, D., Flytzanis, C., Kudryavtsev, I. A., Yazeva, T. V., Rodina, A. V. and Efros, A. L. (1993). Absorption and intensity-dependent photoluminescence measurements on cdse quantum dots - assignment of the 1st electronic-transitions, *J. Opt. Soc. Am. B* **10**, p. 100.

Ellingson, R. J., Beard, M. C., Johnson, J. C., Yu, P., Micic, O. I., Nozik, A. J., Shabaev, A. and Efros, A. L. (2005). Highly efficient multiple exciton generation in colloidal pbse and pbs quantum dots, *Nano Lett.* **5**, pp. 865 – 871.

Ellingson, R. J., Blackburn, J. L., Yu, P., Rumbles, G., Micic, O. I. and Nozik, A. J. (2002). Excitation energy dependent efficiency of charge carrier relaxation and photoluminescence in colloidal inp quantum dots, *J. Phys. Chem. B* **106**, p. 7758.

Exner, P. (2005). Sufficient conditions for the anti-zero effect, *J. Phys. A - Math. Gen.* **38**, p. L449.

Fan, Y., Goldsmith, B. R. and Collins, P. G. (2005). Identifying and counting point defects in carbon nanotubes, *Nature Materials* **4**, p. 906.

Fantini, C., Jorio, A., Souza, M., Ladeira, L. O., Souza, A. G., Saito, R., Samsonidze, G. G., Dresselhaus, G., Dresselhaus, M. S. and Pimenta, M. A. (2004a). One-dimensional character of combination modes in the resonance raman scattering of carbon nanotubes. *Phys. Rev. Lett.* **93**, 8, p. 087401.

Fantini, C., Jorio, A., Souza, M., Saito, R., Samsonidze, G. G., Dresselhaus, M. S. and Pimenta, M. A. (2005). Steplike dispersion of the intermediate-frequency raman modes in semiconducting and metallic carbon nanotubes. *Phys. Rev. B* **72**, 8, pp. 085446 –.

Fantini, C., Jorio, A., Souza, M., Strano, M. S., Dresselhaus, M. S. and Pimenta, M. A. (2004b). Optical transition energies for carbon nanotubes from resonant raman spectroscopy: environment and temperature effects. *Phys. Rev. Lett.* **93**, 14, p. 147406.

Farahani, J., Pohl, D., Eisler, H. and Hecht, B. (2005). Single quantum dot coupled to a scanning optical antenna: A tunable superemitter, *Phys. Rev. Lett.* **95**, p. 017402.

Fernee, M. J., Jensen, P. and Rubinsztein-Dunlop, H. (2007). Origin of the large homogeneous line widths obtained from strongly quantum confined pbs nanocrystals at room temperature. *J. Phys. Chem. C* **111**, 13, p. 4984.

Ferretti, A., Ruini, A., Molinari, E. and Caldas, M. J. (2003). Electronic properties of polymer crystals: The effect of interchain interactions, *Phys. Rev. Lett.* **90**, 8, p. 086401.

Figge, M. T., Mostovoy, M. and Knoester, J. (2001). Peierls transition with acoustic phonons and solitwistons in carbon nanotubes. *Phys. Rev. Lett.* **86**, 20, p. 4572.

Figge, M. T., Mostovoy, M. and Knoester, J. (2002). Peierls instability due to the interaction of electrons with both acoustic and optical phonons in metallic carbon nanotubes. *Phys. Rev. B* **65**, 12, pp. 125416 – 17.

Frackowiak, E. and Beguin, F. (2001). Carbon materials for the electrochemical storage of energy in capacitors, *Carbon* **39**, p. 937.

Franceschetti, A., An, J. M. and Zunger, A. (2006). Impact ionization can explain carrier multiplication in pbse quantum dots, *Nano Lett.* **6**, pp. 2191–2195.

Franco, I. and Tretiak, S. (2004). Electron-vibrational dynamics of photoexcited polyfluorenes, *J. Am. Chem. Soc.* **126**, 38, pp. 12130 – 12140.

Frey, J. T. and Doren, D. J. (2005). *TubeGen 3.3* (University of Delaware, Newark DE).

Frisch, M. J., Trucks, G. W., Schlegel, H. B., Scuseria, G. E., Robb, M. A., Cheeseman, J. R., Montgomery, J. A., Jr., Vreven, T., Kudin, K. N., Burant, J. C., Millam, J. M., Iyengar, S. S., Tomasi, J., Barone, V., Mennucci, B., Cossi, M., Scalmani, G., Rega, N., Petersson, G. A., Nakatsuji, H., Hada, M., Ehara, M., Toyota, K., Fukuda, R., Hasegawa, J., Ishida, M., Nakajima, T., Honda, Y., Kitao, O., Nakai, H., Klene, M., Li, X., Knox, J. E., Hratchian, H. P., Cross, J. B., Bakken, V., Adamo, C., Jaramillo, J., Gomperts, R., Stratmann, R. E., Yazyev, O., Austin, A. J., Cammi, R., Pomelli, C., Ochterski, J. W., Ayala, P. Y., Morokuma, K., Voth, G. A., Salvador, P., Dannenberg, J. J., Zakrzewski, V. G., Dapprich, S., Daniels, A. D., Strain, M. C., Farkas, O., Malick, D. K., Rabuck, A. D., Raghavachari, K., Foresman, J. B., Ortiz, J. V., Cui, Q., Baboul, A. G., Clifford, S., Cioslowski, J., Stefanov, B. B., Liu, G., Liashenko, A., Piskorz, P., Komaromi, I., Martin, R. L., Fox, D. J., Keith *et al.*, Gaussian 03, Revision D.02, Gaussian, Inc., Wallingford, CT, 2004.

Fu, H. X., Wang, L. W. and Zunger, A. (1998). Applicability of the k center dot p method to the electronic structure of quantum dots. *Phys. Rev. B* **57**, 16, pp. 9971 – 9987.

Furche, F. and Ahlrichs, R. (2002). Adiabatic time-dependent density functional methods for excited state properties, *J. Chem. Phys.* **117**, 16, pp. 7433 – 7447.

Gambetta, A., Manzoni, C., Menna, E., Meneghetti, M., Cerullo, G., Lanzani, G., Tretiak, S., Piryatinski, A., Saxena, A., Martin, R. L. and Bishop, A. R. (2006a). Real time observation of non-linear vibrational dynamics in semiconducting single wall carbon nanotubes, *Nature Phys.* **2**, 9, pp. 515–520.

Gambetta, A., Manzoni, C., Menna, E., Meneghetti, M., Cerullo, G., Lanzani, G., Tretiak, S., Piryatinski, A., Saxena, A., Martin, R. L. and Bishop, A. R. (2006b). Real-time observation of nonlinear coherent phonon dynamics in single-walled carbon nanotubes, *Nature Phys.* **2**, 8, pp. 515 – 20.

Gierschner, J., Cornil, J. and Egelhaaf, H. J. (2007). Optical bandgaps of pi-conjugated organic materials at the polymer limit: Experiment and theory, *Adv. Mat.* **19**, 2, pp. 173–191.

Gontijo, I., Buller, G. S., Massa, J. S., Walker, A. C. and Zaitsev, S. V. (1999). Time-resolved photoluminescence and carrier dynamics in vertically-coupled self-assembled quantum dots. *Jpn. J. Appl. Phys.* **38**, p. 674.

Gorman, J., Hasko, D. G. and Williams, D. A. (2005). Charge-qubit operation of an isolated double quantum dot. *Phys. Rev. Lett.* **95**, p. 090502.

Goupalov, S. V., Satishkumar, B. C. and Doorn, S. K. (2006). Excitation and chirality dependence of the exciton-phonon coupling in carbon nanotubes. *Phys. Rev. B* **73**, 11, pp. 115401 – 1.

Gross, E. K. U., Dobson, J. F. and Petersilka, M. (1996). *Density Functional Theory*, Vol. 181 (Springer, Berlin).

Gruneis, A., Saito, R., Samsonidze, G. G., Kimura, T., Pimenta, M. A., Jorio, A., Filho, A. G. S., Dresselhaus, G. and Dresselhaus, M. S. (2003). Inhomogeneous optical absorption around the k point in graphite and carbon nanotubes. *Phys. Rev. B* **67**, 16, pp. 165402 – 1.

Guo, J., Hasan, S., Javey, A., Bosman, G. and Lundstrom, M. (2005). Assessment of high-frequency performance potential of carbon nanotube transistors, *IEEE transactions on nanotechnology* **4**, p. 715.

Guyot-Sionnest, P., Shim, M., Matranga, C. and Hines, M. (1999). *Phys. Rev. B* **60**, p. R2181.

Guyot-Sionnest, P., Wehrenberg, B. and Yu, D. (2005). Intraband relaxation in cdse nanocrystals and the strong influence of the surface ligands. *J. Chem. Phys.* **123**, p. 074709.

Habenicht, B. F., Craig, C. F. and Prezhdo, O. V. (2006). Time-domain ab initio simulation of electron and hole relaxation dynamics in a single-wall semiconducting carbon nanotube. *Phys. Rev. Lett.* **96**, 18, p. 187401.

Habenicht, B. F., Kamisaka, H., Yamashita, K. and Prezhdo, O. V. (2007). Ab initio study of vibrational dephasing of electronic excitations in semiconducting carbon nanotubes, *Nano Lett.* **7**, p. 3260.

Hagen, A., Steiner, M., Raschke, M. B., Lienau, C., Hertel, T., Qian, H., Meixner, A. J. and Hartschuh, A. (2005). Exponential decay lifetimes of excitons in individual single-walled carbon nanotubes, *Phys. Rev. Lett.* **95**, p. 197401.

Hammes-Schiffer, S. and Tully, J. C. (1994). Proton transfer in solution: Molecular dynamics with quantum transitions, *J. Chem. Phys.* **101**, p. 4657.

Harbold, J. M., Du, H., Krauss, T. D., Cho, K.-S., Murray, C. B. and Wise, F. W. (2005a). Time-resolved intraband relaxation of strongly confined electrons and holes in colloidal pbse nanocrystals. *Phys. Rev. B* **72**, 19, p. 195312.

Harbold, J. M., Du, H., Krauss, T. D., Cho, K.-S., Murray, C. B. and Wise, F. W. (2005b). Time-resolved intraband relaxation of strongly confined electrons and holes in colloidal pbse nanocrystals, *Phys. Rev. B* **72**, p. 195312.

Hartschuh, A., Pedrosa, H. N., Novotny, L. and Krauss, T. D. (2003a). Simultaneous fluorescence and raman scattering from single carbon nanotubes. *Science* **301**, 5638, pp. 1354 – 6.

Hartschuh, A., Pedrosa, H. N., Novotny, L. and Krauss, T. D. (2003b). Simultaneous fluorescence and raman scattering from single carbon nanotubes, *Science* **301**, p. 1354.

Heeger, A. J., Kivelson, S., Schrieffer, J. R. and Su, W. P. (1988). Solitons in conducting polymers, *Rev. Mod. Phys.* **60**, 3, pp. 781–850.

Heitz, R., Veit, M., Kalburge, A., Zie, Q. and Grundmann, M. (1998). *Physica E* **2**, p. 578.

Heitz, R., Veit, M., N.Lebentsov, N., Hoffmann, A. and Bimberg, D. (1997). *Phys. Rev. B* **56**, p. 10435.

Hertel, T. and Moos, G. (2000). Electron-phonon interaction in single-wall carbon nanotubes: A time-domain study, *Phys. Rev. Lett.* **84**, p. 5002.

Hertel, T., Zhu, Z., Crochet, J., McPheeters, C., Ulbrecht, H. and Resasco, D. (2006). Exciton dynamics probed in carbon nanotube suspensions with narrow diameter distribution, *Physica Status Solidi B- Basic Solid State Physics* **243**, p. 3186.

Herzberg, G. (1950). *Molecular Spectra and Molecular Structure. I. Spectra of Diatomic Molecules*, Vol. Second Edition (Van Nostrand Reinhold, New York).

Hirori, H., Matsuda, K., Miyauchi, Y., Maruyama, S. and Kanemitsu, Y. (2006). Exciton localization of single-walled carbon nanotubes revealed by femtosecond excitation correlation spectroscopy, *Phys. Rev. Lett.* **97**, p. 25401.

Htoon, H., O'Connell, M. J., Cox, P. J., Doorn, S. K. and Klimov, V. I. (2004). Low temperature emission spectra of individual single-walled carbon nanotubes: multiplicity of subspecies within single-species nanotube ensembles. *Phys. Rev. Lett.* **93**, 2, p. 027401.

Htoon, H., OConnell, M. J., Doorn, S. K. and Klimov, V. I. (2005). Single carbon nanotubes probed by photoluminescence excitation spectroscopy: The role of phonon-assisted transitions, *Phys. Rev. Lett.* **94**, p. 127403.

Hu, Y. Z., Lindberg, M. and Koch, S. W. (1990). Theory of optically excited intrinsic semiconductor quantum dots. *Phys. Rev. B* **42**, 3, pp. 1713 – 23.

Huang, L., Pedrosa, H. N. and Krauss, T. D. (2004). Ultrafast ground-state recovery of single-walled carbon nanotubes, *Phys. Rev. Lett.* **93**, p. 017403.

Hueso, L. E., Pruneda, J. M., Ferrari, V., Burnell, G., Valdes-Herrera, J. P., Simons, B. D., Littlewood, P. B., Artacho, E., Fert, A. and Mathur, N. D. (2007). Transformation of spin information into large electrical signals using carbon nanotubes. *Nature* **445**, 7126, pp. 410 – 13.

Hugle, S. and Egger, R. (2002). van hove singularities in disordered multichannel quantum wires and nanotubes. *Phys. Rev. B* **66**, 19, pp. 193311 – 1.

Hutchison, G. R., Zhao, Y. J., Delley, B., Freeman, A. J., Ratner, M. A. and Marks, T. J. (2003). Electronic structure of conducting polymers: Limitations of oligomer extrapolation approximations and effects of heteroatoms, *Phys. Rev. B* **68**, 3, p. 035204.

Hwang, H. and Rossky, P. J. (2004). Electronic decoherence induced by intramolecular vibrational motions in a betaine dye molecule, *J. Phys. Chem. B* **108**, p. 6723.

Ispasiou, R. G., Lee, J., Papadimitrakopoulos, F. and Goodson, T. (2001). Surface effects in the fluorescence ultrafast dynamics from cdse nanocrystals, *Chem. Phys. Lett.* **340**, p. 7.

Itoh, T., Nishijima, M., Ekimov, A. I., Gourdon, C., Efros, A. L. and Rosen, M. (1995). Polaron and exciton-phonon complexes in cucl nanocrystals. *Phys. Rev. Let.* **74**, 9, pp. 1645 – 8.

Jarillo-Herrero, P., van Dam, J. A. and Kouwenhoven, L. P. (2006). Quantum supercurrent transistors in carbon nanotubes. *Nature* **439**, 7079, pp. 953 – 6.

Jiang, J., Saito, R., Samsonidze, G. G., Chou, S. G., Jorio, A., Dresselhaus, G. and Dresselhaus, M. S. (2005). Electron-phonon matrix elements in single-wall carbon nanotubes. *Phys. Rev. B* **72**, 23, pp. 235408 – 1.

Jones, M., Engtrakul, C., Metzger, W. K., Ellingson, R. J., Nozik, A. J., Heben, M. J. and Rumbles, G. (2005). Analysis of photoluminescence from solubilized single-walled carbon nanotubes, *Phys. Rev. B* **71**, p. 115426.

Jones, M., Metzger, W. K., McDonald, T. J., Engtrakul, C., Ellingson, R. J., Rumble, G. and Heben, M. J. (2007). Extrinsic and intrinsic effects on the excited-state kinetics of single-walled carbon nanotubes, *Nano. Lett.* **7**, p. 300.

Jung, M. S., Choi, T. L., Joo, W. J., Kim, J. Y., Han, I. T. and Kim, J. M. (2007). Transparent conductive thin films based on chemically assembled single-walled carbon nanotubes, *Synthetic Metals* **157**, p. 997.

Kam, N. W. S., O'Connell, M., Wisdom, J. A. and Dai, H. J. (2005). Carbon nanotubes as multifunctional biological transporters and near-infrared agents for selective cancer cell destruction. *Proc. Natl. Acad. Sci.* **102**, pp. 11600 – 5.

Kamisaka, H., Kilina, S. V., Y., Y. and Prezhdo, O. V. (2006). Ultrafast vibrationally-induced dephasing of electronic excitations in pbse quantum dots, *Nano Lett.* **6**, pp. 2295–2300.

Kamisaka, H., Kilina, S. V., Y., Y. and Prezhdo, O. V. (2008). *Ab initio* study of temperature and pressure dependence of energy and phonon-induced dephasing of electronic excitations in cdse and pbse quantum dots, *J. Phys. Chem. C* **112**, pp. 7800–7808.

Kane, C. L. and Mele, E. J. (2004). Electron interactions and scaling relations for optical excitations in carbon nanotubes. *Phys. Rev. Lett.* **93**, 19, p. 197402.

Kang, I. and Wise, F. W. (1997). Electronic structure and optical properties of pbs and pbse quantum dots, *J. Opt. Soc. Am. B* **14**, p. 1632.

Karaiskaj, D., Mascarenhas, A., Choi, J. H., Graff, R. and Strano, M. S. (2007). Temperature behavior of the photoluminescence decay of semiconducting carbon nanotubes: The effective lifetime, *Phys. Rev. B* **75**, p. 113409.

Kasuya, A., Sivamohan, R., Barnakov, Y. A., Dmitruk, I. M., Nirasawa, T., Romanyuk, V. R., Kumar, V., Mamykin, S. V., Tohji, K., Jeyadevan, B., Shinoda, K., Kudo, T., Terasaki, O., Liu, Z., Belosludov, R. V., Sundararajan, V. and Kawazoe, Y. (2004). Ultra-stable nanoparticles of cdse revealed from mass spectrometry. *Nature Mat.* **3**, 2, pp. 99 – 102.

Kayanuma, Y. and Momiji, H. (1990). Incomplete confinement of electrons and holes in microcrystals. *Phys. Rev. B* **41**, 14, pp. 10261 – 3.

Kilina, S. and Tretiak, S. (2007). Excitonic and vibrational properties of single-walled semiconducting carbon nanotubes, *Adv. Funct. Mater.* **17**, pp. 3405 – 3420.

Kilina, S., Tretiak, S., Doorn, S. K., Luo, Z., Papadimitrakopoulos, F., Piryatinski, A., Saxena, A. and Bishop, A. R. (2008). Cross-polarized excitons of carbon nanotubes, *Proc. Nat. Acad. Sci.* **105**, pp. 6797 – 6802.

Kilina, S. V., Craig, C. F., Kilin, D. S. and Prezhdo, O. V. (2007). Ab initio time-domain study of phonon-assisted relaxation of charge carriers in a pbse quantum dot. *J. Phys Chem. C* **111**, 12, p. 4871.

Kim, S. Y. and Hammes-Schiffer, S. (2006). Hybrid quantum/classical molecular dynamics for a proton transfer reaction coupled to a dissipative bath, *J. Chem. Phys.* **124**, p. 244102.

Klessinger, M. and Michl, J. (1995). *Excited States And Photochemistry Of Organic Molecules* (VCH, New York).

Klimov, V., Bolivar, P. H. and Kurz, H. (1995). Hot-phonon effects in femtosecond luminescence spectra of electron-hole plasmas in cds. *Phys. Rev. B* **52**, 7, pp. 4728 – 31.

Klimov, V. I. (2000). Optical nonlinearities and ultrafast carrier, *J. Phys. Chem. B* **104**, p. 6112.

Klimov, V. I. (2004). *Semiconductor and Metal Nanocrystals: Synthesis and Electronic and Optical Properties* (Marcel Dekker, New York).

Klimov, V. I. (2006). Mechanisms for photogeneration and recombination of multiexcitons in semiconductor nanocrystals: Implications for lasing and solar energy conversion. *J. Phys. Chem. B* **110**, 34, p. 16827.

Klimov, V. I., Ivanov, S. A., Nanda, J., Achermann, M., Bezel, I., McGuire, J. A. and Piryatinski, A. (2007). Single-exciton optical gain in semiconductor nanocrystals. *Nature* **447**, 7143, pp. 441 – 6.

Klimov, V. I. and McBranch, D. W. (1998). Femtosecond 1p-to-1s electron relaxation in strongly confined semiconductor nanocrystals, *Phys. Rev. Lett.* **80**, p. 4028.

Klimov, V. I., Mikhailovsky, A. A., McBranch, D. W., Leatherdale, C. A. and Bawendi, M. G. (2000a). Mechanisms for intraband energy relaxation in semiconductor quantum dots: the role of electron-hole interactions, *Phys. Rev. B* **61**, p. R13349.

Klimov, V. I., Mikhailvsky, A. A., Xu, S., V., M. A., Hollingsworth, J. A., Leatherdale, C. A., Eisler, M.-J. and Bawendi, M. G. (2000b). Optical gain and stimulated emission in nanocrystal quantum dots, *Science* **290**, pp. 314–317.

Klimov, V. I., Schwartz, C. J., McBranch, D. W., Leatherdale, C. A. and Bawendi, M. G. (1999). Ultrafast dynamics of inter- and intraband transitions in semiconductor nanocrystals: implications for quantum-dot lasers, *Phys. Rev. B* **60**, p. R2177.

Kondov, I., Kleinekathofer, U. and Schreiber, M. (2003). Stochastic unraveling of redfield master equations and its application to electron transfer problems, *J. Chem. Phys.* **119**, p. 6635.

Kong, J., Franklin, N. R., Zhou, C., Chapline, M. G., Peng, S., Cho, K. and Dai, H. (2000). Nanotube molecular wires as chemical sensors. *Science* **287**, 5453, pp. 622 – 5.

Korovyanko, O. J., Sheng, C. X., Vardeny, Z. V., Dalton, A. B. and Baughman, R. H. (2004a). Ultrafast spectroscopy of excitons in single-walled carbon nanotubes. *Phys. Rev. Lett.* **92**, 1, p. 017403.

Korovyanko, O. J., Sheng, C.-X., Vardeny, Z. V., Dalton, A. B. and Baughman, R. H. (2004b). Ultrafast spectroscopy of excitons in single-walled carbon nanotubes, *Phys. Rev. Lett.* **92**, p. 017403.

Krasheninnikov, A. V. and Banhart, F. (2007). Engineering of nanostructured carbon materials with electron or ion beams. *Nature Mat.* **6**, 10, pp. 723 – 33.

Krauss, T. D. and Wise, F. W. (1997a). Coherent acoustic phonons in a semiconductor quantum dot. *Phys. Rev. Lett.* **79**, 25, pp. 5102 – 5.

Krauss, T. D. and Wise, F. W. (1997b). Raman-scattering study of exciton-phonon coupling in pbs nanocrystals. *Phys. Rev. B* **55**, 15, pp. 9860 – 5.

Kresse, G. and Furthmüller, J. (1996a). *Comput. Mater. Sci.* **6**, pp. 15–50.

Kresse, G. and Furthmüller, J. (1996b). Efficient iterative schemes for ab initio total-energy calculations using a plane-wave basis set, *Phys. Rev. B* **54**, p. 11169.

Kresse, G. and Hafner, J. (1994). Ab initio molecular-dynamics simulation of the liquid-metal wamorphous-semiconductor transition in germanium, *Phys. Rev. B* **49**, p. 14251.

Krupke, R., Hennrich, F., von Lohneysen, H. and Kappes, M. M. (2003). Separation of metallic from semiconducting single-walled carbon nanotubes. *Science* **301**, 5631, pp. 344 – 7.

Kuhn, H. (1948a). Elektronengasmodell zur quantitativen deutung der lichtabsorption von organischen farbstoffen .1. *Helv. Chim. Acta* **31**, 6, pp. 1441 – 1455.

Kuhn, H. (1948b). Free electron model for absorption spectra of organic dyes, *J. Chem. Phys.* **16**, 8, pp. 840 – 841.

Kuhn, H. (1949). A quantum-mechanical theory of light absorption of organic dyes and similar compounds, *J. Chem. Phys.* **17**, pp. 1198 – 1211.

Kuno, M., Fromm, D. P., Hamann, H. F., Gallagher, A. and Nesbitt, D. J. (2001). On/off fluorescence intermittency of single semiconductor quantum dots. *J. Chem. Phys.* **115**, 2, p. 1028.

Lany, S., Osorio-Guillen, J. and Zunger, A. (2007). Origins of the doping asymmetry in oxides: hole doping in nio versus electron doping in zno. *Phys. Rev. B* **75**, 24, pp. 241203 – 4.

Lanzani, G., Cerullo, G., Gambetta, A., Manzoni, C., Menna, E. and Meneghetti, M. (2005). Exciton relaxation in single wall carbon nanotubes, *Synth. Met.* **155**, p. 246.

Larsen, R. E. and Schwartz, B. J. (2006). Nonadiabatic molecular dynamics simulations of correlated electrons in solution. 2. a prediction for the observation of hydrated dielectrons with pump-probe spectroscopy, *J. Phys. Chem. B* **110**, p. 9692.

Ledebo, L. A. and Ridley, B. K. (1982). On the position of energy levels related to transition-metal impurities in iii-v semiconductors. *J. Phys. C* **15**, 27, p. L961.

Lee, J. U. (2005). Photovoltaic effect in ideal carbon nanotube diodes, *Phys. Rev. Let.* **87**, p. 073101.

Lee, J. Y., Connor, S. T., Cui, Y. and Peumans, P. (2008). Solution processed metal nanowire mesh transparent electrodes, *Nano Lett.* **8**, p. 689.

Lefebvre, J., Austing, D. G., Bond, J. and Finnie, P. (2006). Photoluminescence imaging of suspended single-walled carbon nanotubes, *Nano Lett.* **6**, p. 1603.

Lefebvre, J. and Finnie, P. (2007). Polarized photoluminescence excitation spectroscopy of single-walled carbon nanotubes, *Phys. Rev. Lett.* **98**, p. 167406.

Lefebvre, J., Finnie, P. and Homma, Y. (2004a). Temperature-dependent photoluminescence from single-walled carbon nanotubes, *Physical Review B* **70**, p. 045419.

Lefebvre, J., Finnie, P. and Homma, Y. (2004b). Temperature-dependent photoluminescence from single-walled carbon nanotubes, *Phys. Rev. B.* **70**, p. 045419.

Lefebvre, J., Homma, Y. and Finnie, P. (2003). Bright band gap photoluminescence from unprocessed single-walled carbon nanotubes, *Phys. Rev. Lett.* **90**, p. 217401.

LeRoy, B. J., Lemay, S. G., Kong, J. and C, C. D. (2004a). Electrical generation and absorption of phonons in carbon nanotubes, *Nature* **432**, p. 371.

LeRoy, B. J., Lemay, S. G., Kong, J. and Dekker, C. (2004b). Electrical generation and absorption of phonons in carbon nanotubes, *Nature* **432**, 7015, pp. 371 – 374.

Li, X.-Q., Nakayama, H. and Arakawa, Y. (1999). *Jpn. J. Appl. Phys.* **38**, p. 473.

Li, X. S., Tully, J. C., Schlegel, H. B. and Frisch, M. J. (2006). Ab initio ehrenfest dynamics, *J. Chem. Phys.* **123**, p. 084106.

Li, Z. M., Tang, Z. K., Liu, H. J., Wang, N., Chan, C. T., Saito, R., Okada, S., Li, G. D., Chen, J. S., Nagasawa, N. and Tsuda, S. (2001). Polarized absorption spectra of single-walled 4 aa carbon nanotubes aligned in channels of an alposub 4-5 single crystal. *Phys. Rev. Lett.* **87**, 12, pp. 127401 – 4.

Liljerothos, P., van Emmichoven, P. A. Z., Hickey, S. G., Weller, H., Grandidier, B., Allan, G. and Vanmaekelbergh, D. (2005). Density of states measured by scannin-tunneling spectroscopy sheds new light on the optical transition in pbse nanocrystals, *Phys. Rev. Lett.* **95**, p. 086801.

Lim, Y. S., Yee, K. J., Kim, J. H., Haroz, E. H., Shaver, J., Kono, J., Doorn, S. K., Hauge, R. H. and Smalley, R. E. (2006). Coherent lattice vibrations in single-walled carbon nanotubes, *Nano Lett.* **6**, 12, pp. 2696 – 2700.

Linderberg, J., Jorgensen, P., Oddershede, J. and Ratner, M. (1972). Self-consistent polarization propagator approximation as a modified random phase method. *J. Chem. Phys.* **56**, 12, pp. 6213 – 19.

Linderberg, J. and Öhrn, Y. (1973). *Propagators in Quantum Chemistry* (Academic Press, London).

Liu, C. and Cheng, H. M. (2005). Carbon nanotubes for clean energy applications, *Journal of Physics D* **38**, p. 231.

Lockwood, D. M., Cheng, Y. K. and Rossky, P. J. (2001). Electronic decoherence for electron transfer in blue copper proteins, *Chem. Phys. Lett.* **345**, p. 159.

Lowisch, M., Rabe, M., Kreller, F. and Henneberger, F. (1999). *Appl. Phys. Lett* **74**, p. 2489.

Luis, A. (2003). Zeno and anti-zeno effects in two-level systems, *Phys. Rev. A* **67**, p. 062113.

Luis, E. F., Torres, F. and Roche, S. (2006). Inelastic quantum transport and peierls-like mechanism in carbon nanotubes. *Phys. Rev. Lett.* **97**, 7, pp. 076804 – 4.

Luo, Z., Papadimitrakopoulos, F. and Doorn, S. (2008). Bundling effects on the intensities of second-order Raman modes in semiconducting single-walled carbon nanotubes, *Phys. Rev. B* **77**, p. 035421.

Luo, Z. T., Papadimitrakopoulos, F. and Doorn, S. K. (2007). Intermediate-frequency raman modes for the lower optical transitions of semiconducting single-walled carbon nanotubes. *Phys. Rev. B* **75**, 20, pp. 205438 –.

Ma, Y. Z., Stenger, J., Zimmermann, J., Bachilo, S. M., Smalley, R. E., Weisman, R. B. and Fleming, G. R. (2004). Ultrafast carrier dynamics in single-walled carbon nanotubes probed by femtosecond spectroscopy, *J. Chem. Phys.* **120**, p. 3368.

Ma, Y.-Z., Valkunas, L., Bachilo, S. M. and Fleming, G. R. (2005a). Exciton binding energy in semiconducting single-walled carbon nanotubes, *J. Phys. Chem. B* **109**, p. 15671.

Ma, Y.-Z., Valkunas, L., Bachilo, S. M. and Fleming, G. R. (2006). Temperature effects on femtosecond transient absorption kinetics of semiconducting single-walled carbon nanotubes, *Phys. Chem. Chem. Phys.* **8**, p. 5689.

Ma, Y.-Z., Valkunas, L., Dexheimer, S. L., Bachilo, S. M. and Fleming, G. R. (2005b). Femtosecond spectroscopy of optical excitations in single-walled carbon nanotubes: Evidence for exciton-exciton annihilation, *Phys. Rev. Lett.* **94**, 15, p. 157402.

Ma, Y. Z., Valkunas, L., Dexheimer, S. L., Bachilo, S. M. and Fleming, G. R. (2005c). Femtosecond spectroscopy of optical excitations in single-walled carbon nanotubes: Evidence for exciton-exciton annihilation, *Phys. Rev.*

Lett. **94**, p. 157402.

Machon, M., Reich, S., Telg, H., Maultzsch, J., Ordejon, P. and Thomsen, C. (2005). Strength of radial breathing mode in single-walled carbon nanotubes. *Phys. Rev. B* **71**, 3, pp. 35416 – 1.

Machon, M., Reich, S. and Thomsen, C. (2006). Strong electron-phonon coupling of the high-energy modes of carbon nanotubes, *Phys. Rev. B* **74**, 20, pp. 205423 –.

Maniscalco, S., Piilo, J. and Suominen, K. A. (2006). Zeno and anti-zeno effects for quantum brownian motion, *Phys. Rev. Lett.* **97**, p. 130402.

Manzoni, C., Gambetta, A., Menna, E., Meneghetti, M., Lanzani, G. and Cerullo, G. (2005b). Intersubband exciton relaxation dynamics in single-walled carbon nanotubes, *Phys. Rev. Lett.* **94**, 20, pp. 1 – 4.

Manzoni, C., Gambetta, A., Menna, E., Meneghetti, M., Lanzani, G. and Cerullo, G. (2005a). Ultrafast spectroscopy of excitons in single-walled carbon nanotubes, *Phys. Rev. Lett.* **94**, p. 207401.

Maroto, A., Balasubramanian, K., Burghard, M. and Kern, K. (2007). Functionalized metallic carbon nanotube devices for ph sensing. *Chem. Phys. Chem.* **8**, 2, pp. 220 – 3.

Marques, M. A. L. and Gross, E. K. U. (2004). Time-dependent density functional theory, *Annu. Rev. Phys. Chem.* **55**, p. 427.

Mason, N., Biercuk, M. J. and Marcus, C. M. (2004). Local gate control of a carbon nanotube double quantum dot. *Science* **303**, 5658, pp. 655 – 8.

Maultzsch, J., Pomraenke, R., Reich, S., Chang, E., Prezzi, D., Ruini, A., Molinari, E., Strano, M. S., Thomsen, C. and Lienau, C. (2005). Exciton binding energies in carbon nanotubes from two-photon photoluminescence. *Phys. Rev. B* **72**, 24, pp. 241402 – 1.

Metzger, W. K., McDonald, T. J., Engtrakul, C., Blackburn, J. L., Scholes, G. D., Rumbles, G. and Heben, M. J. (2007). Temperature-dependent excitonic decay and multiple states in single-wall carbon nanotubes, *J. Phys. Chem. C.* **111**, p. 3601.

Misewich, J. A., Martel, R., Avouris, P., Tsang, J. C., Heinze, S. and Tersoff, J. (2003). Electrically induced optical emission from a carbon nanotube fet. *Science* **300**, 5620, pp. 783 – 786.

Mittleman, D. M., Schoenlein, R. W., Shiand, J. J., Colvin, V. L., Alivisatos, A. P. and Shank, C. V. (1994). Quantum size dependence of femtosecond electronic dephasing and vibrational dynamics in cdse nanocrystals. *Phys. Rev. B* **49**, 20, p. 14435.

Miyauchi, Y., Oba, M. and Maruyama, S. (2006). Cross-polarized optical absorption of single-walled nanotubes by polarized photoluminescence excitation spectroscopy. *Phys. Rev. B* **74**, 20, pp. 205440 – 1.

Mohamed, M. B., Burda, C. and El-Sayed, M. A. (2001). Shape dependent ultrafast relaxation dynamics in nanocrystals: nanorods vs. nanodots, *Nano Lett.* **1**, p. 589.

Morello, G., Giorgi, M. D., Kudera, S., Manna, L., Cingolani, R. and Anni, M. (2007). Temperature and size dependence of nonradiative relaxation and exciton-phonon coupling in colloidal cdte quantum dots. *J. Phys. Chem. C*

111, 16, pp. 5846 – 9.

Mortimer, I. B. and Nicholas, R. J. (2007). Role of bright and dark excitons in the temperature-dependent photoluminescence of carbon nanotubes, *Phys. Rev. Lett.* **98**, p. 027404.

Mukai, K. and Sugawara, M. (1998). *Jpn. J. Appl. Phys.* **37**, p. 5451.

Mukamel, S. (1995). *Principles of Nonlinear Optical Spectroscopy* (Oxford University Press, New York).

Mukamel, S., Tretiak, S., Wagersreiter, T. and Chernyak, V. (1997). Electronic coherence and collective optical excitations of conjugated molecules, *Science* **277**, 5327, pp. 781–787.

Muljarov, E. A., Takagahara, T. and Zimmermann, R. (2005). Phonon-induced exciton dephasing in quantum dot molecules. *Phys. Rev. Lett.* **95**, 17, p. 177405.

Murdin, B. N., Hollingworth, A. R., Kamal-Saadi, M., Kotitschke, R. T. and Cielsla, C. N. (1999). *Phys. Rev. B* **59**, p. R7817.

Murphy, J. E., Beard, M. C., Norman, A. G., Ahrenkiel, S. P., Johnson, J. C., Yu, P., Mii, O. I., Ellingson, R. J. and Nozik, A. J. (2006). Pbte colloidal nanocrystals: Synthesis, characterization, and multiple exciton generation, *J. Am. Chem. Soc.* **128**, pp. 3241–3247.

Murray, C. B., Norris, D. J. and Bawendi, M. G. (1993). Synthesis and characterization of nearly monodisperse cde (e = s, se, te) semiconductor nanocrystallites. *J. Am. Chem. Soc.* **115**, 19, p. 8706.

Nair, G. and Bawendi, M. G. (2007). Carrier multiplication yields of cdse and cdte nanocrystals by transient photoluminescence spectroscopy. *Phys. Rev. B* **76**, 8, pp. 081304 –.

Nasibulin, A. G., Pikhitsa, P. V., Jiang, H., Brown, D. P., Krasheninnikov, A. V., Anisimov, A. S., Queipo, P., Moisala, A., Gonzalez, D., Lientschnig, G., Hassanien, A., Shandakov, S. D., Lolli, G., Resasco, D. E., Choi, M., Tomanek, D. and Kauppinen, E. I. (2007). A novel hybrid carbon material, *Nature Nanotech.* **2**, 3, pp. 156 – 61.

Negele, J. W. (1982). The mean-field theory of nuclear structure and dynamics, *Rev. Mod. Phys.* **54**, p. 913.

Nozik, A. J. (2001). Spectroscopy and hot electron relaxation dynamics in semiconductor quantum wells and quantum dots, *Annu. Rev. Phys. Chem.* **52**, p. 193.

O'Connell, M. J., Bachilo, S. M., Huffman, C. B., Moore, V. C., Strano, M. S., Haroz, E. H., Rialon, K. L., Boul, P. J., Noon, W. H., Kittrell, C., Ma, J. P., Hauge, R. H., Weisman, R. B. and Smalley, R. E. (2002). Band gap fluorescence from individual single-walled carbon nanotubes, *Science* **297**, p. 787.

O'Connell, M. J., Eibergen, E. E. and Doorn, S. K. (2005). Chiral selectivity in the charge-transfer bleaching of single-walled carbon-nanotube spectra, *Nature Mat.* **4**, 5, pp. 412 – 418.

Onida, G., Reining, L. and Rubio, A. (2002). Electronic excitations: density-functional versus many-body green's-function approaches, *Rev. Mod. Phys.* **74**, 2, pp. 601 – 659.

Oregan, B., Gratzel, M. and Fitzmaurice, D. (1991). Optical electrochemistry .1. steady-state spectroscopy of conduction-band electrons in a metal-oxide semiconductor electrode. *Chem. Physi. Lett.* **183**, 1-2, pp. 89 – 93.

Oron-Carl, M., Hennrich, F., Kappes, M. M., Lohneysen, H. V. and Krupke, R. (2005). On the electron-phonon coupling of individual single-walled carbon nanotubes, *Nano Lett.* **5**, p. 1761.

Ostojic, G. N., Zaric, S., Kono, J., Strano, M. S., Moore, V. C., Hauge, R. H. and Smalley, R. E. (2004). Interband recombination dynamics in resonantly excited single-walled carbon nanotubes, *Phys. Rev. Lett.* **92**, p. 117402.

Ouyang, M. and Awschalom, D. D. (2003). Coherent spin transfer between molecularly bridged quantum dots, *Science* **301**, p. 1074.

Parahdekar, P. V. and Tully, J. C. (2005). Mixed quantum-classical equilibrium, *J. Chem. Phys.* **122**, p. 094102.

Peierls, R. E. (1955). *Quantum Theory of Solids* (Clarendon, Oxford, UK).

Perdew, J. P. (1991). *Electronic Structure of Solids*, p. ziesche and h. eschrig edn. (Akademie Verlag, Berlin).

Perebeinos, V., Tersoff, J. and Avouris, P. (2004). Scaling of excitons in carbon nanotubes, *Phys. Rev. Lett.* **92**, 25, p. 257402.

Perebeinos, V., Tersoff, J. and Avouris, P. (2005a). Effect of exciton-phonon coupling in the calculated optical absorption of carbon nanotubes. *Phys. Rev. Lett.* **94**, 2, p. 027402.

Perebeinos, V., Tersoff, J. and Avouris, P. (2005b). Electron-phonon interaction and transport in semiconducting carbon nanotubes. *Phys. Rev. Lett.* **94**, 8, p. 086802.

Perebeinos, V., Tersoff, J. and Avouris, P. (2005c). Radiative lifetime of excitons in carbon nanotubes, *Nano Lett.* **5**, 12, pp. 2495 – 2499.

Perebeinos, V., Tersoff, J. and Avouris, P. (2006). Mobility in semiconducting carbon nanotubes at finite carrier density, *Nano Lett.* **6**, 2, pp. 205 – 208.

Perebinos, V., Tersoff, J. and Avouris, P. (2005a). Effect of exciton-phonon coupling in the calculated optical absorption of carbon nanotubes, *Phys. Rev. Lett.* **94**, p. 027402.

Perebinos, V., Tersoff, J. and Avouris, P. (2005b). Electron-phonon interaction and transport in semiconducting carbon nanotubes, *Phys. Rev. Lett.* **94**, p. 086802.

Peterson, J. J. and Krauss, T. D. (2006). Fluorescence spectroscopy of single lead sulfide quantum dots. *Nano Lett.* **6**, 3, p. 510.

Petta, J. R., Johnson, A. C., Taylor, J. M., Laird, E. A., Yacoby, A., Lukin, M. D., Marcus, C. M., Hanson, M. P. and Gossard, A. C. (2005). Coherent manipulation of coupled electron spins in semiconductor quantum dots. *Science* **309**, p. 2180.

Pines, D. and Bohm, D. (1952). A collective description of electron interactions .2. collective vs individual particle aspects of the interactions, *Phys. Rev.* **85**, 2, pp. 338 – 353.

Plentz, F., Ribeiro, H. B., Jorio, A., Strano, M. S. and Pimenta, M. A. (2005a). Direct experimental evidence of exciton-phonon bound states in carbon nanotubes. *Phys. Rev. Lett.* **95**, 24, p. 247401.

Plentz, F., Ribeiro, H. B., Jorio, A., Strano, M. S. and Pimenta, M. A. (2005b). Direct experimental evidence of exciton-phonon bound states in carbon nanotubes. *Phys. Rev. Lett.* **95**, 24, pp. 247401 – 4.

Pollock, E. L. and Koch, S. W. (1991). Path-integral study of excitons and biexcitons in semiconductor quantum dots. *J. Chem. Phys.* **94**, 10, pp. 6776 – 81.

Postma, H. W. C., Teepen, T., Yao, Z., Grifoni, M. and Dekker, C. (2001). Carbon nanotube single-electron transistors at room temperature. *Science* **293**, 5527, pp. 76 – 79.

Prabhu, S. S., Vengurlekar, A. S., Roy, S. K. and Shah, J. (1995). Nonequilibrium dynamics of hot carriers and hot phonons in cdse and gaas. *Phys. Rev. B* **51**, 20, pp. 14233 – 14246.

Prezhdo, O. V. (2000). Quantum anti-zeno acceleration of a chemical reaction, *Phys. Rev. Lett.* **85**, p. 4413.

Prezhdo, O. V. and Brooksby, C. (2001). Quantum backreaction through the bohmian particle. *Phys. Rev. Lett.* **86**, 15, pp. 3215 – 19.

Prezhdo, O. V. and Rossky, P. J. (1997a). Evaluation of quantum transition rates from quantum-classical molecular dynamics simulations, *J. Chem. Phys.* **15**, p. 5863.

Prezhdo, O. V. and Rossky, P. J. (1997b). Mean-field molecular dynamics with surface hopping. *J. Chem. Phys.* **107**, 3, pp. 825 – 34.

Prezhdo, O. V. and Rossky, P. J. (1998). Quantum decoherence and short time solvent response, *Phys. Rev. Lett.* **81**, p. 5294.

Pulci, O., Onida, G., Sole, R. D. and Reining, L. (1998). Ab initio calculation of self-energy effects on optical properties of gaas(110), *Phys. Rev. Lett.* **81**, p. 5374.

Puzder, A., Williamson, A. J., Gygi, F. and Galli, G. (2004). Self-healing of cdse nanoctystals: first-principles calculations, *Phys. Rev. Lett.* **92**, p. 217401.

Qiu, X. H., Freitag, M., Perebeinos, V. and Avouris, P. (2005). Photoconductivity spectra of single-carbon nanotubes: Implications on the nature of their excited states, *Nano Lett.* **5**, 4, pp. 749 – 752.

Rafailov, P. M., Maultzsch, J., Thomsen, C. and Kataura, H. (2005). Electrochemical switching of the peierls-like transition in metallic single-walled carbon nanotubes, *Phys. Rev. B* **72**, 4, pp. 045411 –.

Regan, B. C., Aloni, S., Ritchie, R. O., Dahmen, U. and Zettl, A. (2004). Carbon nanotubes as nanoscale mass conveyors. *Nature* **428**, 6986, pp. 924 – 7.

Reining, L., Rubio, A., Vast, N. and Marinopoulos, A. G. (2003). Optical and loss spectra of carbon nanotubes: Depolarization effects and intertube interactions, *Phys. Rev. Lett.* **91**, 4, pp. 046402/1 – 046402/4.

Rettrup, S. (1982). An iterative method for calculating several of the extreme eigensolutions of large real nonsymmetric matrices, *J. Comp. Phys.* **45**, 1, pp. 100–107.

Ridley, B. K. (1982). *Quantum Process in semiconducors* (Clarendon Press, Oxford).

Ring, P. and Schuck, P. (1980). *The Nuclear Many-Body Problem* (Springer-Verlag, New York).

Roche, S., Jiang, J., Triozon, F. and Saito, R. (2005a). Quantum dephasing in carbon nanotubes due to electron-phonon coupling. *Phys. Rev. Lett.* **95**, 7, pp. 076803 – 4.

Roche, S., Jiang, J., Triozon, F. and Saito, R. (2005b). Quantum dephasing in carbon nanotubes due to electron-phonon coupling, *Phys. Rev. Lett.* **95**, p. 076803.

Saito, R., Dresselhaus, G. and Dresselhaus, M. S. (1998a). *Physical properties of carbon nanotubes* (Imperial College Press, London, UK).

Saito, R., Dresselhaus, G. and Dresselhaus, M. S. (1998b). *Physical Properties of Carbon Nanotubes* (Imperial College Press, London).

Salvetat, J. P., Bhattacharyya, S. and Pipes, R. B. (2006). Pogress on mechanics of carbon nanotubes and derived materials, *Journal of Nanoscience and Nanotechnology* **6**, p. 1857.

Samuel, I. D. W., Ledoux, I., Dhenaut, C., Zyss, J., Fox, H. H., Schrock, R. R. and Silbey, R. J. (1994). Saturation of cubic optical nonlinearity in long-chain polyene oligomers, *Science* **265**, 5175, pp. 1070–1072.

Schaller, R. D., Agranovich, V. M. and Klimov, V. I. (2005a). High-efficiency carrier multiplication through direct photogeneration of multi-excitons via virtual single-exciton states, *Nat. Phys.* **1**, p. 189.

Schaller, R. D., Agranovich, V. M. and Klimov, V. I. (2005b). High-efficiency carrier multiplication through direct photogeneration of multi-excitons via virtual single-exciton states, *Nature Phys.* **1**, p. 189.

Schaller, R. D. and Klimov, V. I. (2004). High efficiency carrier multiplication in pbse nanocrystals: Implications for solar energy conversion, *Phys. Rev. Lett.* **92**, p. 186601.

Schaller, R. D., Petruska, M. A. and Klimov, V. I. (2005c). Effect of electronic structure on carrier multiplication efficiency: Comparative study of pbse and cdse nanocrystals, *Appl. Phys. Lett.* **87**, p. 253102.

Schaller, R. D., Pietryga, J. M., Goupalov, S. V., Petruska, M. A., Ivanov, S. A. and Klimov, V. I. (2005d). Breaking the phonon bottleneck in semiconductor nanocrystals via multiphonon emission induced by intrinsic nonadiabatic interactions, *Phys. Rev. Lett.* **95**, p. 196401.

Schaller, R. D., Sykora, M., Pietryga, J. M. and Klimov, V. I. (2006). Seven excitons at a cost of one: Redefining the limits for conversion efficiency of photons into charge carriers. *Nano Lett.* **6**, 3, p. 424.

Scheibner, R., Buhmann, H., Reuter, D., Kiselev, M. and Molenkamp, L. (2005). Thermopower of a kondo spin-correlated quantum dot, *Phys. Rev. Lett.* **95**, p. 176602.

Schleser, R., Ihn, T., Ruh, E., Ensslin, K., Tews, M., Pfannkuche, D., Driscoll, D. and Gossard, A. (2005). Cotunneling-mediated transport through excited states in the coulomb-blockade regime, *Phys. Rev. Lett.* **94**, p. 206805.

Scholes, G., Tretiak, S., McDonald, T., Metzger, W., Engtrakul, C., Rumbles, G. and Heben, M. (2007). Low-lying exciton states determine the photophysics of semiconducting single wall carbon nanotubes, *J. Phys. Chem. C* **111**, p. 11139.

Scholes, G. D. and Rumbles, G. (2006). Excitons in nanoscale systems, *Nature*

Mat. **5**, 9, pp. 683 – 696.

Schwartz, B. J., Bittner, E. R., Prezhdo, O. V. and Rossky, P. J. (1996). Quantum decoherence and the isotope effect in condensed phase nonadiabatic molecular dynamics simulations, *J. Chem. Phys.* **104**, p. 5942.

Seferyan, H. Y., Nasr, M. B., Senekerimyan, V., Zadoyan, R., Collins, P. and Apkarian, V. A. (2006). Transient grating measurements of excitonic dynamics in single-walled carbon nanotubes: The dark excitonic bottleneck, *Nano Lett.* **6**, p. 1757.

Semenov, Y. G. and Kim, K. W. (2004). Phonon-mediated electron-spin phase diffusion in a quantum dot. *Phys. Rev. Lett.* **92**, 2, p. 026601.

Sercel, P. C. and Vahala, K. J. (1990). Analytical formalism for determining quantum-wire and quantum-dot band structure in the multiband envelope-function approximation. *Phys. Rev. B* **42**, 6, pp. 3690 – 710.

Set, S. Y., Yaguchi, H., Tanaka, Y. and Jablonski, M. (2004a). Laser mode locking using a saturable absorber incorporating carbon nanotubes. *J. Lightwave Tech.* **22**, 1, pp. 51 – 6.

Set, S. Y., Yaguchi, H., Tanaka, Y. and Jablonski, M. (2004b). Ultrafast fiber pulsed lasers incorporating carbon nanotubes. *IEEE J. Sel. Top. Quant. El.* **10**, 1, pp. 137 – 46.

Sewall, S. L., Cooney, R. R., Anderson, K. E. H., Dias, E. A. and Kambhampati, P. (2006). State-to-state exciton dynamics in semiconductor quantum dots, *Phys. Rev. B* **74**, 23, 235328, doi:10.1103/PhysRevB.74.235328, URL `http://link.aps.org/abstract/PRB/v74/e235328`.

Sfeir, M. Y., Beetz, T., Wang, F., Huang, L. M., Huang, X. M. H., Huang, M. Y., Hone, J., O'Brien, S., Misewich, J. A., Heinz, T. F., Wu, L. J., Zhu, Y. M. and Brus, L. E. (2006). Optical spectroscopy of individual single-walled carbon nanotubes of defined chiral structure. *Science* **312**, 5773, pp. 554 – 6.

Shaver, J., Kono, J., Portugall, O., Krstic, V., Rikken, G., Miyauchi, Y., Maruyama, S. and Perebeinos, V. (2007). Magnetic brightening of carbon nanotube photoluminescence through symmetry breaking. *Nano Lett.* **7**, 7, pp. 1851 – 1855.

Shimizu, K. T., Neuhauser, R. G., Leatherdale, C. A., Empedocles, S. A., Woo, W. K. and Bawendi, M. G. (2001). Blinking statistics in single semiconductor nanocrystal quantum dots. *Phys. Rev. B* **63**, 20, p. 205316.

Shreve, A. P., Haroz, E. H., Bachilo, S. M., Weisman, R. B., Tretiak, S., Kilina, S. and Doorn, S. K. (2007a). Determination of exciton-phonon coupling elements in single-walled carbon nanotubes by raman overtone analysis, *Phys. Rev. Lett.* **98**, 3, 037405.

Shreve, A. P., Haroz, E. H., Bachilo, S. M., Weisman, R. B., Tretiak, S., Kilina, S. and Doorn, S. K. (2007b). Determination of exciton-phonon coupling elements in single-walled carbon nanotubes by Raman overtone analysis, *Phys. Rev. Lett.* **98**, 3, p. 037405.

Simpson, W. T. (1955). Resonance force theory of carotenoid pigments, *J. Am. Chem. Soc.* **77**, 23, pp. 6164 – 6168.

Skinner, J. L. (1988). The theory of pure dephasing in crystals, *Ann. Rev. Phys.*

Chem. **39**, p. 463.

Smalley, R. E., Li, Y., Moore, V. C., Price, B. K., Jr., R. C., Schmidt, H. K., Hauge, R. H., Barron, A. R. and Tour, J. M. (2006). Single wall carbon nanotube amplificiation: en route to a type-specific growth mechanism, *J. Am. Chem. Soc.* **128**, p. 15824.

Snow, E. S., Perkins, F. K., Houser, E. J., Badescu, S. C. and Reinecke, T. L. (2005). Chemical detection with a single-walled carbon nanotube capacitor. *Science* **307**, 5717, pp. 1942 – 1945.

Sosnowski, T. S., Norris, T. B., Jiang, H., Singh, J., Kamath, K. and Bhattacharya, P. (1998). *Phys. Rev. B* **57**, p. R9423.

Spataru, C. D., Ismail-Beigi, S., Benedict, L. X. and Louie, S. G. (2004). Excitonic effects and optical spectra of single-walled carbon nanotubes, *Phys. Rev. Lett.* **92**, 7, p. 774021.

Spataru, C. D., Ismail-Beigi, S., Capaz, R. B. and Louie, S. G. (2005a). Theory and ab initio calculation of radiative lifetime of excitons in semiconducting carbon nanotubes, *Phys. Rev. Lett.* **95**, p. 247402.

Spataru, C. D., Ismail-Beigi, S., Capaz, R. B. and Louie, S. G. (2005b). Theory and ab initio calculation of radiative lifetime of excitons in semiconducting carbon nanotubes, *Phys. Rev. Lett.* **95**, p. 247402.

Sternberg, M., Curtiss, L. A., Gruen, D. M., Kedziora, G., Horner, D. A., Redfern, P. C. and Zapol, P. (2006). Carbon ad-dimer defects in carbon nanotubes, *Phys. Rev. Lett.* **96**, p. 075506.

Stewart, J. J. P. (2000). *MOPAC 2002* (Schrödinger Inc. and Fujitsu Limited, Portland, OR 97201).

Stratmann, R. E., Scuseria, G. E. and Frisch, M. J. (1998). An efficient implementation of time-dependent density-functional theory for the calculation of excitation energies of large molecules, *J. Chem. Phys.* **109**, 19, pp. 8218–8224.

Streed, E. W., Mun, J., Boyd, M., Campbell, G. K., Medley, P., Ketterle, W. and Pritchard, D. E. (2006). Continuous and pulsed quantum zeno effect, *Phys. Rev. Lett.* **97**, p. 260402.

Sugawara, M., Mukai, K. and Shoji, H. (1997). *Appl. Phys. Lett* **71**, p. 2791.

Szabo, A. and Ostlund, N. S. (1989). *Modern Quantum Chemistry: Introduction to Advanced Electronic Structure Theory* (McGraw-Hill, New York).

Takagahara, T. (1993). Effects of dielectric confinement and electron-hole exchange interaction on excitonic states in semiconductor quantum dots. *Phys. Rev. B* **47**, 8, pp. 4569 – 84.

Talapin, D. V. and Murray, C. B. (2005). Pbse nanocrystal solids for n- and p-channel thin film field-effect transistors. *Science* **310**, 5745, p. 86.

Tang, Z. K., Zhang, L. Y., Wang, N., Zhang, X. X., Wen, G. H., Li, G. D., Wang, J. N., Chan, C. T. and Sheng, P. (2001). Superconductivity in 4 angstrom single-walled carbon nanotubes, *Science* **292**, p. 2462.

Tans, S. J., Devoret, M. H., Dai, H. J., Thess, A., Smalley, R. E., Geerligs, L. J. and Dekker, C. (1997). Individual single-wall carbon nanotubes as quantum wires, *Nature* **386**, p. 474.

Tans, S. J., Verschueren, A. R. M. and Dekker, C. (1998). Carbon nanotubes as quantum wires, *Nature* **393**, p. 49.

Telg, H., Maultzsch, J., Reich, S., Hennrich, F. and Thomsen, C. (2004). Chirality distribution and transition energies of carbon nanotubes. *Phys. Rev. Lett.* **93**, 17, pp. 177401 – 4.

Telg, H., Maultzsch, J., Reich, S. and Thomsen, C. (2006). Resonant-raman intensities and transition energies of the esub 11 transition in carbon nanotubes. *Phys. Rev. B* **74**, 11, pp. 115415 – 1.

Terabe, K., Hasegawa, T., Nakayama, T. and Aono, M. (2005). Quantized conductance atomic switch, *Nature* **433**, 7021, pp. 47 – 50.

Thouless, D. J. (1972). *The Quantum Mechanics Of Many-Body Systems* (Academic Press, New York).

Tretiak, S. (2007a). Triplet state absorption in carbon nanotubes: A td-dft study. *Nano Let.* **7**, 8, pp. 2201 – 2206.

Tretiak, S. (2007b). Triplet state absorption in carbon nanotubes: A td-dft study, *Nano Lett.* **7**, p. 2201.

Tretiak, S., Igumenshchev, K. and Chernyak, V. (2005). Exciton sizes of conducting polymers predicted by time-dependent density functional theory. *Phys. Rev. B* **71**, 3, p. 33201.

Tretiak, S., Kilina, S., A.Piryatinski, Saxena, A., Martin, M. L. and Bishop, A. R. (2007a). Excitons and peierls distortion in conjugated carbon nanotubes, *Nano Lett.* **7**, p. 86.

Tretiak, S., Kilina, S., Piryatinski, A., Saxena, A., Martin, R. L. and Bishop, A. R. (2007b). Excitons and peierls distortion in conjugated carbon nanotubes, *Nano Lett.* **7**, 1, pp. 86 – 92.

Tretiak, S. and Mukamel, S. (2002). Density matrix analysis and simulation of electronic excitations in conjugated and aggregated molecules, *Chem. Rev.* **102**, 9, pp. 3171–3212.

Tretiak, S., Saxena, A., Martin, R. L. and Bishop, A. R. (2000a). Ceo/semiempirical calculations of uv-visible spectra in conjugated molecules, *Chem. Phys. Lett.* **331**, 5-6, pp. 561–568.

Tretiak, S., Saxena, A., Martin, R. L. and Bishop, A. R. (2000b). Interchain electronic excitations in poly(phenylenevinylene) (ppv) aggregates, *J. Phys. Chem. B* **104**, 30, pp. 7029–7037.

Tretiak, S., Saxena, A., Martin, R. L. and Bishop, A. R. (2002). Conformational dynamics of photoexcited conjugated molecules, *Phys. Rev. Lett.* **89**, 9, p. 097402.

Tretiak, S., Saxena, A., Martin, R. L. and Bishop, A. R. (2003). Photoexcited breathers in conjugated polyenes: An excited-state molecular dynamics study, *Proc. Nat. Acad. Sci. USA* **100**, 5, pp. 2185 – 2190.

Trickey, S. B. (Ed.) (1990). *Density-Functional Theory of Many-Fermion Systems* (Academic Press, Boston).

Tully, J. C. (1990). Molecular dynamics with electronic transitions, *J. Chem. Phys.* **93**, p. 1061.

Uryu, S. and Ando, T. (2006). Exciton absorption of perpendicularly polarized light in carbon nanotubes, *Phys. Rev. B* **74**, p. 155411.

Vanderbilt, D. (1990). Soft self-consistent pseudopotentials in a generalized eigenvalue formalism, *Phys. Rev. B* **41**, p. 7892.

Vandescuren, M., Amara, H., Langlet, R. and Lambin, P. (2007). Characterization of single-walled carbon nanotubes containing defects from their local vibrational densities of states, *Carbon* **45**, p. 359.

Wang, F., Dukovic, G., Brus, L. E. and Heinz, T. F. (2004). Time-resolved fluorescence of carbon nanotubes and its implication for radiative lifetimes, *Phys. Rev. Lett.* **92**, p. 177401.

Wang, F., Dukovic, G., Brus, L. E. and Heinz, T. F. (2005a). The optical resonances in carbon nanotubes arise from excitons, *Science* **308**, 5723, pp. 838 – 841.

Wang, F., Dukovic, G., Brus, L. E. and Heinz, T. F. (2005b). The optical resonances in carbon nanotubes arise from excitons, *Science* **308**, p. 838.

Wang, J. (2005). Carbon-nanotube based electrochemical biosensors: a review, *Electroanalysis* **17**, p. 7.

Wang, L. W., Califano, M., Zunger, A. and Franceschetti, A. (2003). Pseudopotential theory of auger processes in cdse quantum dots. *Phys. Rev. Lett.* **91**, 5, pp. 056404 –.

Wang, W. and Zunger, A. (1995). Pbse spectra, *Phys. Rev. B* **51**, p. 17398.

Wang, Z., Pedrosa, H., Krauss, T. and Rothberg, L. (2006a). Determination of the exciton binding energy in single-walled carbon nanotubes, *Phys. Rev. Lett.* **96**, 4, p. 115415.

Wang, Z. D., Zhao, H. B. and Mazumdar, S. (2006b). Quantitative calculations of the excitonic energy spectra of semiconducting single-walled carbon nanotubes within a pi-electron model. *Phys.l Rev. B* **74**, 19, pp. 195406 –.

Wang, Z. D., Zhao, H. B. and Mazumdar, S. (2007). π-electron theory of transverse optical excitons in semiconducting single-walled carbon nanotubes, *Phys. Rev. B* **76**, 11, p. 115431.

Wehrenberg, B. L., Wang, C. and Guyot-Sionnest, P. (2002). Interband and intraband optical studies of pbse colloidal quantum dots. *J. Phys. Chem. B* **106**, 41, pp. 10634 – 40.

Weisman, R. B. and Bachilo, S. M. (2003). Dependence of optical transition energies on structure for single-walled carbon nanotubes in aqueous suspension: an empirical kataura plot. *Nano Lett.* **3**, 9, pp. 1235 – 8.

Wilson-Rae, I., Zoller, P. and Imamoglu, A. (2004). Laser cooling of a nanomechanical resonator mode to its quantum ground state, *Phys. Rev. Lett.* **92**, p. 075507.

Wise, F. W. (2000). Lead salt quantum dots: The limit of strong quantum confinement. *Acc. Chem. Res.* **33**, 11, p. 773.

Woggon, U., Giessen, H., Gindele, F., Wind, O., Fluegel, B. and Peyghambarian, N. (1996). Ultrafast energy relaxation in quantum dots. *Phys. Rev. B* **54**, 24, pp. 17681 – 90.

Wu, C., Malinin, S. V., Tretiak, S. and Chernyak, V. Y. (2006). Exciton scattering and localization in branched dendrimeric structures, *Nature Phys.* **2**, 9, pp. 631 – 635.

Xu, S., Mikhailovsky, A. A., Hollingsworth, J. A. and Klimov, V. I. (2002). Hole intraband relaxation in strongly confined quantum dots: Revisiting the phonon bottleneck problem. *Phys. Rev. B* **65**, 4, pp. 045319 – 5.

Yanagi, K., Iakoubovskii, K., Matsui, H., Matsuzaki, H., Okamoto, H., Miyata,

Y., Maniwa, Y., Kazaoui, S., Minami, N. and Kataura, H. (2007). Photosensitive function of encapsulated dye in carbon nanotubes. *J. Am. Chem. Soc.* **129**, 16, pp. 4992 – 4997.

Yarotski, D., Kilina, S., Talin, A., Tretiak, S., Balatsky, A. and Taylor, A. (2008, in press). Scanning tunneling microscopy of DNA-wrapped carbon nanotubes, *Nano Lett.* .

Yin, Y., Vamivakas, A. N., Walsh, A. G., Cronin, S. B., Unlu, M. S., Goldberg, B. B. and Swan, A. K. (2007). Optical determination of electron-phonon coupling in carbon nanotubes, *Phys. Rev. Lett.* **98**, 3, 037404.

Yoffe, A. D. (2001). Semiconductor quantum dots and related systems: Electronic, optical, luminescence and related properties of low dimensional systems. *Adv. Phys.* **50**, 1, pp. 1 – 208.

Yu, H., Lycett, S., Roberts, C. and Murray, R. (1996). *Appl. Phys. Lett* **69**, p. 4087.

Yu, P., Nedeljkovic, J. M., Ahrenkiel, P. A., Ellingson, R. J. and Nozik, A. J. (2004). Size dependent femtosecond electron cooling dynamics in cdse quantum rods, *Nano Lett.* **4**, p. 1089.

Zamkov, M., Woody, N., Shan, B., Chang, Z. and Richard, P. (2005). Lifetime of charge carriers in multiwalled nanotubes, *Phys. Rev. Lett.* **94**, p. 056803.

Zaric, S., Ostojic, G. N., Shaver, J., Kono, J., Portugall, O., Frings, P. H., Rikken, G. L. J. A., Furis, M., Crooker, S. A., Wei, X., Moore, V. C., Hauge, R. H. and Smalley, R. E. (2006). Excitons in carbon nanotubes with broken time-reversal symmetry, *Phys. Rev. Lett.* **96**, p. 016406.

Zerner, M. C. (1996). *ZINDO, A semiempirical quantum chemistry program* (Quantum Theory Project, University of Florida, Gainesville, FL).

Zhang, M., Fang, S. L., Zakhidov, A. A., Lee, S. B., Aliev, A. E., Williams, C. D., Atkinson, K. R. and Baughman, R. H. (2005). Strong, transparent, multifunctional, carbon nanotube sheets, *Science* **309**, 5738, pp. 1215 – 1219.

Zhao, H., Mazumdar, S., Sheng, C. X., Tong, M. and Vardeny, Z. V. (2006). Photophysics of excitons in quasi-one-dimensional organic semiconductors: single-walled carbon nanotubes and pi -conjugated polymers. *Phys. Rev. B* **73**, 7, pp. 75403 – 1.

Zhao, H. B. and Mazumdar, S. (2004a). Electron-electron interaction effects on the optical excitations of semiconducting single-walled carbon nanotubes, *Phys. Rev. Lett.* **93**, 15, p. 157402.

Zhao, H. B. and Mazumdar, S. (2004b). Electron-electron interaction effects on the optical excitations of semiconducting single-walled carbon nanotubes, *Phys. Rev. Lett.* **93**, p. 157402.

Zheng, M., Jagota, A., Semke, E. D., Diner, B. A., Mclean, R. S., Lustig, S. R., Richardson, R. E. and Tassi, N. G. (2003a). DNA-assisted dispersion and separation of carbon nanotubes, *Nature Mat.* **2**, 5, pp. 338 – 342.

Zheng, M., Jagota, A., Strano, M. S., Santos, A. P., Barone, P., Chou, S. G., Diner, B. A., Dresselhaus, M. S., McLean, R. S., Onoa, G. B., Samsonidze, G. G., Semke, E. D., Usrey, M. and Walls, D. J. (2003b). Structure-based carbon nanotube sorting by sequence-dependent dna assembly, *Science*

302, 5650, pp. 1545 – 1548.

Zhou, Z. Y., Brus, L. and Friesner, R. (2003). Electronic structure and luminescence of 1.1-and 1.4-nm silicon nanocrystals: Oxide shell versus hydrogen passivation, *Nano Lett.* **3**, p. 163.

Zhou, Z. Y., Steigerwald, M., Hybertsen, M., Brus, L. and Friesner, R. A. (2004). Electronic structure of tubular aromatic molecules derived from the metallic (5,5) armchair single wall carbon nanotube, *J. Am. Chem. Soc.* **126**, 11, pp. 3597 – 3607.

Zunger, A. (2001). Pseudopotential theory of semiconductor quantum dots, *Phys. Stat. Sol. B* **224**, p. 727.

Index